THE QUANTUM MINDSET
IN A NUTSHELL

THE QUANTUM MINDSET IN A NUTSHELL

What They Won't Teach You in School. That Could Change Your Life

RICK THOMPSON

BLASTOFF

BlastOff Publishing LLC.
Powered by H.A.B.I.T

Contents

Part 1: Embracing a New World View

Part 2: Making Quantum Changes in Your Life

THE INTRODUCTION YOU SHOULDN'T SKIP

It's safe to say that we don't know what we don't know, but as long as we know that there are things that we don't know, we have the opportunity for growth in all areas of life. If we embrace the energy of curiosity and choose to live our life with an open mind, we are able to expand our world view and evolve *to the next level...*

What's the next level?

Famous Apple co-founder, Steve Jobs, was rumored to dislike focus groups. He stated, "A lot of times, people don't know what they want until you show it to them." Similarly, automobile pioneer, Henry Ford, famously stated, "If I had asked people what they wanted, they would have said faster horses."

This raises the question: What's the next big thing to come that we may not be aware of yet? Is it another revolutionary invention like the internet, a new communication device that's better than a cell phone, or maybe it's a faster transportation system, something more efficient than an airplane? What if it's better than all the aforementioned combined, something with the power to affect all aspects of life on this planet... a revolution of the mind and a new way of thinking called *The Quantum Mindset?*

Quantum-mind thinking is not only the next level of thinking, it's the next stage in our human evolution. Soon the world will realize that reality isn't real until we give our energy to it; our conscious mind observes and chooses what to focus on by exercising its free will, while our subconscious mind is the powerhouse constructing the reality we choose to create based on our beliefs and choices. The quantum mindset doesn't embody this as a new age teaching. This is cosmic law that we've all inherited at birth but have forgotten as we were led astray.

The best thing about achieving the quantum mindset is that you do not need to acquire a college degree or have any scholarly training. In fact, you don't need to learn much at all. What you will need to do is unlearn that which no longer serves your highest good and open your mind to the infinite possibilities that surround you.

The only thing you will need to "learn" isn't something that needs to be learned *per se*, but more so perceived. The only thing holding you back from *accessing* your quantum mindset are your unspoken rules. I later explain and refer to these unspoken rules as "negative one-liners." These rules are beliefs you acquired starting from adolescence throughout your adulthood, and most are likely unbeknownst to you at the moment.

Become aware of the things you tell yourself every day. What are your unconscious beliefs? Did they serve their purpose then, and do they serve your purpose now? Do these rules make you happy, or do these rules stress you out? Most importantly, where and when did you set those rules for yourself, and why do you still believe them?

Have you ever wondered why certain things are the way they are? I often share the following story of five monkeys and a banana as an example.

In an experiment, five monkeys were placed in a cage together, with a banana hanging on a rope just outside their reach. Researchers placed a step ladder in the middle of the cage which would allow the monkeys to reach the banana. Whenever one of the monkeys attempted to climb for the banana, ALL five monkeys were sprayed with freezing cold water just as soon as the monkey reached for it.

Researchers repeated this several times until the monkeys caught on and made the association between reaching for the banana and the collective punishment of being sprayed with freezing cold water. Soon, there was no longer a need for the water. None of the monkeys would even attempt to reach for the banana due to their fear of being sprayed again.

Next, they replaced one of the five monkeys with a new monkey. Of course, the new monkey, not aware of the cold-water punishment, tried to reach for the banana and immediately the other four monkeys attacked him. Eventually, he stopped trying to reach the banana.

One by one, the original monkeys who had first experienced the ice-water treatment were replaced by a new monkey. The same cycle repeated as the new monkeys were attacked by the other monkeys until they quit trying to reach the banana.

Eventually, five new monkeys occupied the cage, and none of the original monkeys who had experienced the icy water treatment remained. The researchers introduced a new monkey to the cage with the others who had never been sprayed. When this monkey tried to reach for the banana, all five monkeys attacked him, without knowing why they were doing what they were doing.

What did this study teach us? Even though none of these monkeys knew about the group punishment of being sprayed with cold water, along the way they learned that reaching for the banana is forbidden. They became the enforcers of this *rule* without knowing its origins or purpose.

Cultural habits, parental teachings, group beliefs, and religious rituals are created over time and passed to the next generation. Often these practices are ingrained and shared without anyone knowing why, when, or where it started. After all, most people blindly follow the behavior of others in a group just to be accepted, while those who have already accepted the new culture and rules will fight to keep them intact.

So, the next time someone says, "This is just how we do things around here," or, "This is how it's always been," or, "It is what it is," ask the question, "*Why?*" You'll be surprised to hear that many people don't know why they do what they do or believe what they believe.

The truth is: the only thing we get to do in life is decide what *our* rules are, what we focus on and what we think about. The problem is that so much information is presented to us that we're told we need to believe, but the information just isn't true. The world changes and things evolve, yet we don't realize it. We unconsciously adopt old beliefs, then pass them on or miscommunicate them, because we misunderstood or misinterpreted them. This fatal coding error now becomes our "stake in the ground," cemented in our rules and keeping us from going anywhere.

WE BECOME WHAT WE BELIEVE

There's a story of the elephant and the rope, where a baby elephant is tied to a stake with a rope. He pulls and pulls and pulls, but he can't budge the stake in the ground. He finally gives up, believing he'll never be able to pull out the stake or break the rope. A few years go by, and the elephant becomes a 4,000-pound giant, still confined only by a little, itty-bitty rope and stake. But the elephant believes he can't break the rope, and remains caught, even though he is big and strong enough to easily break free.

In the old science-based mindset, people believed many things were quite impossible. For instance, they thought the power of the mind to influence matter was nonexistent. Because of this belief, everyone operated within the confines of a purely materialistic universe. Then along came the quantum gang, who disproved that limiting belief, and now we're able to envision and act on expanded beliefs of who we are and what we can accomplish in the world. As kids, we were programmed with one-liner beliefs that often seriously limited our full potential. Now we're purposefully expanding our beliefs.

You may be asking yourself, "OK, so how do I crack the code and change my rules so I can begin accessing my quantum mindset?" Well, grab a notepad and a pen and let me show you how. We'll start off slow then pick up the pace as we go. Just remember to have an open mind, and don't worry ... your brain won't fall out (wink wink).

BE LIKE WATER

"You must be shapeless, formless, like water. When you pour water in a cup, it becomes the cup. When you pour water in a bottle, it becomes the bottle. When you pour water in a teapot, it becomes the teapot. Water can drip and it can crash. Become like water my friend."

—Bruce Lee

Almost everyone has heard the phrase, "Is the glass half empty or half full?" This question relates to how you perceive life; do you focus on the good or do you focus on the bad? Do you focus on abundance or do you focus on scarcity? Before this cognitive reflection, we just saw a glass with water, with no forward-thinking, just identifying black and white. But with a quantum mindset, we no longer ask if the glass is half empty or half full. Instead, we ask, "Was the water poured in, or was it poured out? And what can I do with this half glass of water?"

You will soon see life as infinite possibilities and probabilities that will benefit you and everyone around you.

Water itself is an amazing element. All life started from water. Water holds a vibration better than just about any other element. Therefore, the possibility of what we could do with water is endless.

The study headed by Dr. Masaru Emoto is a great example. In this study, scientists took two glasses of water and emotionally showered one glass of water with TLC (Tender Loving Care) by verbally and repeatedly telling the water it was loved, with all sorts of positive words. Meanwhile, the scientists isolated the other glass of water and did the opposite by saying horrible things to it. Then they froze the two glasses of water. Afterwards, the water that had been showered with positive words of love had frozen into patterns of beautiful snowflakes and geometric shapes, which amazed the scientists. When they examined the other glass, it had noticeably abnormal and non-symmetrical shapes that were far less recognizable compared to the beautiful snowflakes in the previous glass. Below are actual images of the results when this study was performed, where the glasses of water were exposed to positive and negative communication, frozen, then examined under a high-powered microscope.

LOVE PLUS GRATITUDE

YOU MAKE ME SICK, I WILL KILL YOU

Photo taken from www.SpiritScience.com

Water is the most receptive of the five elements in Taoism (wood, fire, earth, metal, water), and it's a scientific fact that our bodies are made up of more than 75 percent water. If mere negative thoughts and discouraging words can have such a significant impact on water, since our bodies are made up of 75 plus percent water, imagine what our thoughts and negative words can do to us.

NOTE: The practice of *Quantum Water Jumping* has been going viral on YouTube recently. You can check out this new rapid-manifestation phenomenon on my website: TheQuantumMindset.com.

WHAT IS THE QUANTUM MINDSET?

It's no longer a secret—human society is currently in deep transformation regarding how we habitually see ourselves, how we engage with the world, and how we accomplish our goals. Similar to the previous mind-jolt 500 years ago when people began to realize the world was not a flat surface but a sphere hurtling through space, recent scientific discoveries have exploded our classical scientific assumptions about the nature of the universe.

We all know that what we believe to be true tends to be what we manifest in our lives. Everything we create in our outer world begins as a dream, an idea, a vision that grows outward from our imagination into material expression. This means that if we want to change our lives for the better, we first need to hold an inner mindset that reflects the most up-to-date quantum understanding of how the world works.

Much of our world is rapidly being transformed by science and technological advances that most of us haven't kept up with the new vision of reality that quantum physics has revealed. We're stuck trying to operate in a rapidly-changing world using old assumptions and unrealistic expectations. The result is that we're not succeeding at our highest potential, because we're not taking advantage of the many breakthroughs in science, technology, and psychology that can expand and enlighten our personal world view.

To correct this dilemma, we need to take *action*—we need to shift into a new mindset that empowers us to leap into a more productive and fulfilling life experience. We need to consciously let go of old out-moded beliefs and attitudes about how the world functions and embrace a quite remarkable new vision of who we are, and how we can best interact with the world to get what we truly want.

As an example of how great this leap into a quantum mindset can be, consider the quantum-physics discovery that everything around us that we assumed to be "dead," inanimate matter is actually vibrantly alive at subatomic levels, empowered with remarkable energy, intelligence, attraction, and engagement. When this new understanding of reality begins to sink into our world view, it carries the power to transform everything in our lives.

In the vast macro-dimension of astrophysics, revolutionary new discoveries shatter our old beliefs and present an expanded vision of the universe that directly impacts our sense of who we are and how our lives are evolving. For instance, we were taught that the universe expands at a steady speed, but now we know the whole universe is actually accelerating. Even at the cosmic level, things are speeding up, just as our personal lives seem to be going faster and faster. When we take in this new information about the universe, we can break out of "future shock" and expand our world view to embrace the expansive nature of the reality we live in.

Likewise in our relationships, in our ability to attract to us the people and things and situations we desire, potent discoveries are being made about the underlying forces of nature that hold everything together. By better understanding these fundamental forces, we can learn how to work with them to transform the way we attract a better life for ourselves and our loved ones.

Most of us don't have the time, interest, or the mental muscle to stay on top of the deluge of weird reports in the news about mind-boggling nano-technologies, quirky quark particles, and ten-dimensional string-theories. Meanwhile, quantum science makes vast exciting leaps forward, regarding what's inside black holes and how we might fly off to Mars or someday become superheroes by linking our brains with computers.

But is all this remarkable science and new technology actually improving our lives? Sometimes it seems just the opposite—we're being bombarded with non-stop disastrous news about global warming, deadly pollution, rampant over-population, greed and crime and tribalism and all the rest. As science explores the frayed edges of the universe, we're struggling to avoid anxiety and depression, control our work situation, put bread on the table, and survive yet another challenging (or boring) day.

And yet ... on the brighter side of all this, even as we focus on our immediate personal dilemmas, we can also sense that we're all caught up together in a historic movement that carries the potential to transform our human experience on this planet. A breathtaking new world view is emerging that holds the promise of transforming our personal and shared potential.

In this book, we're going to explore how you can use the insights of quantum science and psychology to your practical advantage. We're going to show you how to expand your own mindset to be mentally prepared to participate more fully in this unique new phase of human evolution.

Furthermore, this transition into the quantum mindset doesn't have to be hard or painful. My job is to make the new science clear and personal to your life. Your job will be to take in new information and ideas and allow your world view (perspective) to evolve as you embrace this new quantum vision of reality. As your traditional view of reality that you were taught in school explodes into a vastly greater realm of possibility, you'll become empowered to dream bigger dreams and use new creative powers to manifest those dreams.

THE PROCESS OF MANIFESTATION

For decades now, many people have been exploring how new breakthroughs in quantum science and positive psychology can help us more successfully manifest what we want in our lives. Along the way, people have indulged in pseudoscience, and in reaction, many people are now put off by any suggestion that we can merge physics and psychology to develop new methods for consciously manifesting our dreams in outer reality.

In this book, I will do my best to keep one foot solidly grounded in the new physics and the other foot equally grounded in positive psychology. But, as many quantum scientists assert, there's much more to reality than the perceived material plane of our senses and the ground we stand on. Just because something can't be experimentally isolated and proven using current materialist-based sensing instruments doesn't mean it doesn't exist. In fact quantum science is based on the acceptance of invisible elements of reality.

QUANTUM IDENTITY

We all seem to be playing God in our power to manifest what we want in life—first by thinking, imagining, or dreaming a great plan or idea, and then by working to manifest that thought or vision into three-dimensional material existence. How do we take an initial great idea and move through all the steps needed to manifest an imagined inner vision out into the world? And how can quantum science shed light on how to manifest at higher levels?

The answer to this question is the basis of this book: how to achieve the quantum mindset. The secret lies within the workings of the conscious mind (observation/focus and initial decision-making) and the subconscious mind (beliefs and automatic decision-making).

Quantum science shows us that particles of our reality do not become particles until they are observed by consciousness. Meaning, it's our own conscious observation of energy waves that changes the wave into a particle. This proves that we do indeed influence our surroundings simply by giving our most valuable commodity to it, our focus! This quantum revelation gives merit to the saying, *where attention goes, energy flows.*

The subconscious mind is the true believer because it acts like a child. Yes, its powerful, but it's also gullible and suggestible. Like a child, it sees and hears EVERYTHING, even when you don't think it does. It's also very black and white and believes or doesn't believe using no real logic, just a decisive decision. The energy put out by the subconscious mind is so fast and strong that our conscious mind could never keep up. Our subconscious is making decisions for us and creating our reality before the conscious mind has time to even think about it. The key is learning how to align your conscious mind and your subconscious mind, so they work *for* you instead of *against* you.

These strange phenomenon of the subconscious mind and manifestation through observation is something we'll continue to delve into in upcoming chapters. It's simpler than you think, and quantum science explains how it all happens.

One of the first steps in higher-level manifesting, as we're exploring throughout this book, is to identify and put aside any ingrained limiting beliefs that deny your personal power to manifest something new. Specifically, we will explore several methods for clearing out old limiting ideas, programming, attitudes, rules, and beliefs that hinder your entry into a highly-creative quantum mindset. Each chapter in this book will offer a practical exercise or experience that you can apply to your current mindset to remove mental and emotional log-jams that block this expansion. The second part of the book focuses intently on the inner-workings of mindset expansion.

Ultimately you'll want to take my ideas and suggestions and make up your own mind about what's real, based on current research and your own inner perspective and experience. Please don't take my word for it. Instead, trust your intuition, insight, and inspiration (the imaginary element), combined with the science of logical deduction. Then you'll be in a position to decide for yourself what to include in your new mindset. And remember, the quantum mindset is not like static—it's always changing as you take in new information and have inner flashes and realizations.

Manifesting anything means dealing with change. A basic rule of quantum mechanics is that change is the primary constant in the universe, so your personal mindset is also something that's constantly emerging. You *can* evolve your world view as we're suggesting. Accepting that change is constant is in itself a quantum leap; go beyond the old notion of static rigid permanence into the realization that in this expanding energetic universe, where everything's in motion, we're constantly evolving into the new.

Within this hopeful sense of who we really are and how the universe functions lies our core promise: that we can now attain greater things in life than before, because we can more consciously imagine and actively materialize our unique vision of the life we want to create. "As above, so below" is an ancient aphorism that still rings true in our scientific era. In the holistic spirit of quantum physics, the laws of creation must be the same for the universe as a whole and for the manifestation of your own personal life vision. Let's dig into what this means.

Part 1:

What Is The Quantum Mindset?

Chapter 1

EVOLUTION OF OUR BRAIN

The brain is an amazing thing, with a vast array of abilities, energies, and powers, most of which is untapped potential. It's also physically astonishing, as it folds on itself many times to fit inside our head. The brain has evolved into three different structures and each stage of structure is associated with its own purpose: the reptilian brain, the mammalian brain, and the human brain (neocortex brain).

Let us begin with the oldest part of the brain, named the reptilian brain because it includes the brainstem and the cerebellum, which are the main structures found in a reptile's brain. This is the part of the brain that keeps us alive every day. As the oldest part of the brain structure, it controls the vital actions that our bodies do automatically, such as breathing, the beating of the heart, and circulating the blood pumping throughout the body. It also regulates our body temperature and is responsible for our balance. It's the part of the brain which is programmed with the basic functions to be a living vessel.

Now we move on to the next level of the brain structure, the mammalian brain; scientifically called the Limbic Brain. This part of the brain is a fascinating and complex part of our basic survival instinct. It's made of the hippocampus, amygdala, and the hypothalamus. It's called the mammalian brain because it emerged from the first mammals, being a mix of our subconscious thinking, instinctive reflexes, and the emotional feeling part of the brain that's intertwined with our physical bodily functions. It's responsible for the physical actions we do, the emotions attached to them, and the good or bad feelings we associate with an experience such as eating, drinking, sensing danger, having sex, and even caring for a loved one.

The mammalian brain holds our memories, thoughts, emotions, sense of pleasure, and our movements. For example, it is responsible for reacting instantly when sensing danger, the sense of pleasure we feel when we eat our favorite food and do the happy-food dance in our seat, or when we are intimate with someone and feel stimulation and satisfaction or dissatisfaction followed by the emotions that bloom from the experience. It's also deeply connected to our sense of smell. Have you ever smelled a scent and instantly you are back in a memory and the feelings attached to that memory? All of these examples are the mammalian brain at work.

Next is the human brain, otherwise known as the neocortex brain. It's the real powerhouse part of the brain because it does the logical and abstract thinking; giving us the ability to think, reason, divide processes, build tools, and come up with means and methods. It also is responsible for our endless learning abilities, our imaginations, and our consciousness. The combination of all these is what helps us evolve and what makes us unique.

All of these brain structures are interconnected and undeniably affect one another. If our reptilian brain ever had an event which affected its automatic functions, such as surviving, then it imprinted the event into our mammalian brain as something dangerous or bad. This affected our human brain by setting limits to its own possibilities to learn, adapt, and imagine new ways of living and even new inventions.

Quick example: A child is swimming near the shore of a beach. While swimming, she sees a small treasure chest under the water. Although it's a little further from her usual comfort zone for swimming, she dives for the treasure box. She can't pick it up easily so she decides she will open it under the water. She goes back up to the surface a few times to get some air. Finally, she opens the latch on the small treasure chest and sees the treasure inside. She reaches for the treasure. Yet, just as her fingers touch it, she is forcibly pulled up by her hair by the hand of her frantic mother (who assumed she was drowning). She opens her mouth automatically, reacting to the unexpected pain of having her hair yanked and it is then that she inhales water and begins to drown as she is brought to the surface. The feeling of drowning due to diving for treasure is so terrifying that the once curious and fearless girl grows up afraid of taking chances to obtain her most desired treasures, her goals.

Scientifically, what happened in the wiring between the three structures of her brain? Her mammalian brain associated taking chances (or going out of her comfort zone) as something dangerous because a function in the reptilian part of her brain, her breathing, was interrupted briefly. This installed a self-limiting belief in her human brain: reaching a goal is not worth taking chances.

SEEING IN A NEW WAY

We tend to see reality based on past experiences and what we expect to see in the future. Our personal awareness is selective and heavily programmed. How we see ourselves and our potential in life determines what we think, imagine, talk about, and do each new day. Developing a quantum mindset means putting aside old ways of seeing things, dismantling outdated beliefs, and embracing a new vision based on the latest clarifications that research provides us.

If seeing is believing, it's important to see the world through the most up-to-date vista that science can provide. You can then integrate these insights into your evolving personal world view. To do this, you'll need to regularly tweak the very foundation of your own consciousness. Your awareness has the inherent ability to expand. When you expand your awareness and align your personal sense of reality with the newest insights from quantum research, you'll position yourself optimally to make the very best of your life.

For decades now, especially since the advent of the computer, startling insights drawn from theories of relativity and quantum physics have challenged our traditional sense of what it means to be conscious beings on this planet. Consciousness itself, long excluded from scientific research and discussions, is finally being considered an integral although mysterious part of physics and psychology. Professors are now insisting that the human mind is flexible and has vast untapped potential. This means you can learn to use your mind in new ways to more effectively attain what you need and desire.

For instance, science revolutionary Sir Isaac Newton laid the foundation of scientific mechanics in the classic Newtonian world view, but the possibility of inner thoughts somehow impacting material reality was considered pure hogwash. Ever since Einstein shook the foundations of Newtonian science, the traditional assumption that your inner thoughts and feelings don't impact the outside world has been openly questioned. Einstein himself dropped the bombshell idea that the mental intent of an experimenter will always influence the outcome of an experiment.

The Nobel-winner Werner Heisenberg went even further, claiming that even in science, a person's inner experience must be considered just as real and valid as any external event. From his lofty academic position in quantum mechanics, Heisenberg postulated the presence of an original non-material cosmic consciousness which was somehow envisioning and manifesting the physical universe. David Bohm, one of the most distinguished theoretical physicists, logically deduced that there must exist what he called a holistic "implicate order" of consciousness from which the explicate order of matter is continually being born.

This might all sound like some vague New Age fantasy, but in fact it's drawn from the newest quantum theories. What's more, it relates directly to how you see yourself, and how you can advance your outside world by changing your inside perspective.

For instance, when you consider your own mind from a quantum perspective and dig down to the smallest material building blocks of the physical brain, you'll find that each neuron in the brain consists of tiny charged particles which are actually not physical at all. Quantum science now insists that the entire brain is made up of non-material energetic coded waves that appear as solid particles only when we disrupt them with an experiment.

Furthermore, quantum theory suggests that the mind is actually an energetic resonance that extends beyond the physical brain, literally connecting you via its wave-like function with the entire universe. The brain itself runs on electricity and generates an electromagnetic force field all around it. We'll see in later chapters how important this is for our interactions with the outside world.

Even as you tentatively entertain such new scientific ideas, you are already actively expanding your mindset to include new possibilities. Once you come to accept an expanded possibility, you'll be empowered to put that expanded vision to use—that's the whole premise of this book. Quantum theories continue to require further development, but we already know this is the general direction the physics world view is moving us in.

What's key, even this early in our discussion, is for you to quietly contemplate such new possibilities and see if they ring true in your own heart and mind. In this light, another new mind-boggling discovery is that both your heart and your skin are deeply entwined neurologically with your brain, and all three work together, along with your other senses, to create your inner experience. So, it's valid to ask not only your brain, but also your heart *and* the feelings in *your* whole body whether a new idea or belief rings true for you. If it does, then pop … your mind just took another quantum leap!

YOU GOTTA HAVE A DREAM

From birth onward, you've been developing your own unique world view—your inner image and model of reality. Even as a baby, you were gobbling new sensory experiences and began to create your mindset to reflect what you were hearing and seeing. Soon, you were learning a language that enabled you to begin thinking about life and making plans to do things out in the world.

All along, you've been growing new neural patterns in your brain coded with the sensory information you were taking in, plus the values and beliefs your culture wanted you to include as the foundation of your world view. By the time you came of age, you possessed a highly-complex inner matrix inside your mind—a unique world view of who you are, how the world around you works, and how you can best interact with the world to create the life you want.

Even now, you're constantly evolving that mindset as you input new experiences and ideas. Your mind is always quite naturally busy creating an internal representation of the world around you, as well as a sense of your own identity and capacity. And here's the basic human situation: you can't really advance in life beyond your own mindset's limits.

In earlier times, when there was less change in the world, a person's childhood mindset or world view could remain mostly intact and unchallenged throughout the span of an entire lifetime. But now, as change accelerates in the world on all fronts, we're being regularly challenged to update our world view, to expand and restructure our personal mindset to more effectively match the emerging scientific and psychological vision of reality.

This human ability and willingness to regularly update our mindset is both a blessing and a curse. The curse is that we have to let go of cherished and secure notions of what life is all about, so we can leap into a new expanded world view. This is often very difficult, especially when our old beliefs are being threatened with upstart ideas about who we are and what life is all about. But the blessing outweighs the curse, because by letting go of the old and embracing the new scientific vision of reality, we can more effectively synch our personal lives with the underlying laws and dynamics of the universe.

Life itself has always been mostly a mystery for human beings. Traditionally people used religion and philosophy, intuition and dreaming, to synch with the subtle unseen dimensions of reality. Long before the advent of modern science, mystics, seers, and spiritual masters would look deeply inward in meditation to directly encounter their own conscious presence, and experience for themselves the "implicate order" and integral wholeness of the universe, often referred to as God, the Tao, Allah, Great Spirit, and so forth.

Before Einstein appeared, traditional Newtonian science actively debunked all internal mental and emotional experience as being non-material and therefore unimportant to science. Any experiential phenomena that couldn't be measured physically with sensitive equipment in an experiment was simply ignored. But Einstein and his gang, and then the new quantum-physics gang, step-by-courageous-step shattered that overly-materialistic notion of reality.

Quantum levels of experimentation have now proven mystics right—an energetic non-material dimension to reality does exist. Furthermore, the thoughts we think, the visions we imagine, the dreams we have, and even the weird psychedelic experiences many people are exploring must all be considered as equally valid and as "real" as a sensory physical experience. Why? Because we are motivated and empowered as much by our inner dreams and imaginations as we are by our external experiences.

It's now time for us as a culture to take seriously the inner, more mysterious realms of consciousness. Even the scientists are saying so, as we'll explore in this book. Openly embracing rather than denying the inherent mysteriousness of life is perhaps the biggest leap of all, and certainly is the most important if your goal is to learn how to manifest your mind's visions into three-dimensional physical reality.

I remember an old musical tune from *South Pacific* by Rogers and Hammerstein that went: "You gotta have a dream. If you don't have a dream, how you gonna have a dream come true?" Because your inner vision is what leads you to manifest your outer world, it's vitally important to honor and employ that inner vision so you can create the life you want.

40 IS THE NEW 20

In an experiment, researchers took older people to a new environment to live for a certain length of time. The entire setting was staged to simulate the 1950s, just like a movie set. In this new environment, the participants were encouraged to live *as if* they were younger, living in the 1950s again.

Before bringing them to their new destination, doctors took their vital signs to get a snapshot of their health and vitality levels. They also tested them after the experiment, and participants had notably started to reverse their aging process while living this more youthful lifestyle. Just by being in an environment where they believed they were living in their younger days it had a profound effect on their physical health and energy levels. This is why the saying goes, *"You don't stop playing because you grow old, you grow old because you stop playing."*

Famous Yogi, Yoganandya, had a profound quote in his book that I'll share. For those of you who don't know of Yoganandya, maybe you have heard of Steve Jobs. Steve Jobs read Yoganandya's book every year, *Autobiography of a Yogi*, and had copies of the book distributed at his funeral.

In this best-selling book, Yoganandya states, *"Environment is stronger than willpower."* He was onto something here. We have all seen the opposite of the 50s study where drug addicts leave their environment and get cleaned up and sober, only to come back to the same setting where their problem started and relapse because their environment overrode their willpower.

IMAGINATION AND REALITY

Until recently, almost all of us were programmed with the cultural prejudice that our inner imaginations weren't really real, that they were mere "figments of our imagination" and not to be taken seriously. But even back in the heyday of Greek civilization, early mathematicians such as Heron of Alexandria were already insisting that to make mathematical sense of the world, there had to be a way to represent "the unseen" non-material dimensions in math formulas.

Humans have always known from the inside-out that there's more to life than meets the eye. Along with our dreams and imaginations and ephemeral flights of fancy, we also have a consciousness that can tune into what we call intuitions, hunches, realizations, and a host of what are labelled mystic experiences. Even five hundred years ago, mathematicians who were trying to make logical sense of reality realized that they had to include non-material dimensions in their calculations.

In 1637, Rene Descartes coined the term "imaginary numbers." Ever since then, mathematicians have been using the "i" symbol in equations to represent unseen dimensions or elements that can't yet be observed or proven, but which simply must be included in math formulas in order to make sense of observed reality.

Einstein knew all about imaginary numbers and unseen levels of reality. Ever since his advent, science has rapidly expanded its prevailing mindset to include both physical material measurable reality, and also the more mysterious non-material dimensions of reality (for instance, dark matter).

Consciousness itself, that non-material and utterly mysterious dynamic that oversees our mindset and resulting thoughts and experiences, is now finally being admitted into formal consideration as a unifying aspect of reality, yet we're still struggling to inject this element into our scientific mindset. By definition, the tools of the experimental method cannot register the presence of consciousness or imagination, intuition or intention on its mechanistic dials, but mathematics and quantum physics insist that the imaginary "i" aspect of reality is of equal importance to the "real" aspects that can be detected physically.

What does this mean to all of us who aren't busy trying to explain the universe in math equations? It means that we can now logically expand our world view, our personal mindset, to include not only the physical sensory stuff of life, but also all our inner realms of consciousness, and we can do this with full scientific approval. If our best scientists must include both imaginary and real numbers to make sense of reality, we can now openly embrace a mindset that includes our sensory experience, and also our deeper inner realms of imagination, intuition, intention, and vision.

Almost all ancient religions are grounded in a universal myth where God creates the universe through either thinking it into being, or imagining and dreaming it into existence. The Christian gospel of Saint John begins with the declaration, "In the beginning was the Word, and the Word was with God, and the Word was God." The Greek term for "Word" in this context, *Logos*, meant "wisdom," a term that includes both reason and intuition. An early theologian noted, "There is one Almighty God, who made all things by His Word, both visible and invisible."

Two thousand years ago in the Judeo-Christian tradition, there was clear mention of a vast unifying force beyond the universe that created the whole universe through employing the Word. Note that a "word" is a verbalized thought expressed through vibration, through waves, and this creation included both the material (visible) and equally the non-material (invisible) aspects of reality. There are parallel creation myths throughout ancient human traditions.

NEW FREEDOM OF CHOICE

One of the main precepts of the quantum revolution is that our whole sense of "freedom of choice" needs to be re-evaluated. In pre-scientific eras, most children were programmed to believe that as humans they were born with free choice, especially in choosing between being bad and being good. Within this theological sense of free will, people had alternatives. They could freely decide to sin or not to sin—to follow the rules or violate them and suffer or enjoy the consequences.

Then in the Newtonian era, children were taught that they lived in a material universe where freedom of choice was exchanged in favor of a predictable world view of cause-and-effect outcomes. This was an age of physical laws, of blunt determinism where the whole universe could be viewed as a linear predictable unfolding. Once a logical flow of events was established, at any point the outcome was already predetermined. There was no free will at all; the mystery of the universe was entirely replaced by scientific analysis.

Next came the theory of relativity to challenge the proclaimed objectivity of the material world view. Einstein concluded that the only constant was the speed of light, and that all observation was subjective, which meant there was no absolute right or wrong. With quantum physics, the whole notion of freedom of choice has returned—because in any given situation or event, there will always be multiple possible outcomes. Probability is the new norm, not linear predictability; we regained free will in a new sense, by having alternate outcomes in any situation.

This new world view offers an escape, for instance, from the assumption that we're all hopeless victims of our childhood beliefs and programming. Now within the quantum mindset, even if our subconscious mind was programmed with a rigid set of limiting rules and fear-based negative beliefs, if they no longer serve us we can choose to discard them. We can consciously act to replace them with more realistic rules and assumptions that serve us better. We'll discuss more about this later.

Quantum theory takes the earlier religious notion of free will another step, transforming traditional "choice" into "engagement" and "connectedness." As we'll explore in depth, the new scientific notion of "entanglement" shows how we're all naturally participating in an inclusive and cohesive larger whole. In fact, it now seems true that every quantum particle, every wave packet in the universe is energetically engaged with all other particles and waves.

This quantum truth is stated in clear terms in the Conway-Kochen theorem: "In pure quantum theory the physical state of the entire universe evolves as one—everything is entangled, and entanglement is not causality, it's connectedness. Both you and the electron are connected with the whole rest of the universe. What you're doing, and what the electron is doing—in fact what every individual thing is doing—is just what the whole universe is doing right now."

The fact that quantum physics takes us beyond free will into a state of universal connectedness is, in my opinion, truly a great leap forward for humankind. We're now at the same time free to explore probabilities and new possibilities, and also free to participate in manifesting the unified happening of each emerging moment in the universe. Through the powers of entanglement and universal interconnectivity, we can finally see ourselves as active participants in the creation of our still-emerging universe—this is truly an exciting expansion of our world view!

As we enter into a quantum mindset, we discover we can set ourselves free to feel and think and act however we want. At the same time, within our quantum mindset, we can imagine and experience being part of the universal whole, participating in harmony with a higher universal will. At first, this might seem to be a paradox, and indeed, it's one well worth contemplating, as its solution lays the inner groundwork for the new quantum consciousness.

DARING TO BELIEVE

This unification of freedom and engagement can become one of your primary powers in life. As you learn to be aware of multiple options and new possibilities and dare to leap into them, you can feel in harmony with your higher calling to unite free will with a sense of harmonious participation in all of creation. This unification of personal will power and a surrender to the higher good will provide you with the greatest manifestation power possible.

For instance, even though perhaps right now you don't have a million dollars (which often represents safety, freedom, power, and so forth), you can employ your freedom of imagination to begin to let yourself feel, think, walk, and act like you're worth a million bucks. If that feeling is strong enough, your beliefs about having a million dollars will begin to shift from unrealistic to realistic to inevitable. Once this happens, your subconscious mind begins to help create this reality for you instead of sabotaging it as your beliefs come in harmony with your wants. The subconscious picks up on all the small things that you say which may prevent you from being in alignment with the belief of having a million dollars. This is why it's so important to be aware of your thoughts and your words; they are manifestations of what you truly believe.

Right now, you're free to imagine and think thoughts that make you feel wealthy, and these invisible thoughts and feelings will begin to change your external life in the direction you choose to grow.

In this new mindset, you can actively align your personal intent with the overall intent of the universe. This isn't just a fanciful New Age notion. Quantum physics, as we'll further clarify step-by-step, verifies that you are an integral connected part of the whole universal. Furthermore, you will be able to observe if your personal ideas and intentions are currently in harmony with the higher universal will. When these two line up—*pow!* You have the power of the universe at your disposal.

> *"Imagination is more important than knowledge. For knowledge is limited, whereas imagination embraces the entire world, stimulating progress, giving birth to evolution."*
> —Albert Einstein

By choosing to improve your inner visions, mental attitudes, and to broadcast positive intentions outward into the physical world, you can rapidly advance both your physical and your emotional reality.

Tentative research at the Princeton PEAR Project has already documented that focused attention can impact a sensitive random-number generator (RNG) at any distance.

Dr. Roger Nelson, who began the study in 1997, created a random number generator (RNG) box, which is a box (the size of two cigarette packs) that he hooked up to his computer. He designed the box to act as a device that flips a coin to land on heads or tails. The theory is that if the coin were flipped one hundred times, it would land 50 percent of the time on heads and 50 percent of the time on tails. Researchers then brought in groups of volunteers who were told to try to influence the ratio of heads to tails and make heads come up 60 percent of the time. It worked! By focusing their mental energy, they changed the outcome.

Other research showed that one person could consciously broadcast a particular word or number into another person's mind at a distance. Such research is difficult and remains controversial because, as Einstein predicted, the intention of the *experimenter* would indeed influence the outcome of the experiment.

More importantly, the proven theories of quantum mechanics show that in each event you move through, there are always multiple choices, and if you just slightly alter your usual habitual choices ten times in a row, you will end up altering the course of your life. Therefore, developing a mindset that stays highly alert to multiple choices is the key to taking charge of the life you want to create.

When you choose to develop an expanded mindset that fully incorporates new scientific insights into your personal world view, you can learn to see yourself and the world around you in a new light and interact with the world in new productive ways. This book is dedicated to helping you take this quantum leap. We're also dedicated to making each step a pleasure to explore!

ILLUSION VERSUS REALITY

We all learned to see through specific cultural filters, and we still see what we *expect* to see. If we can change those expectations, we can then see in a new way and envision fresh new possibilities. Your culturally-programmed mindset unconsciously determines how you see the world and relate to it. If you see the world through a prism of bigotry, prejudice, racial profiling, and fear-based stereotypes, you will view the world quite differently than if you see the world through a prism of equality, acceptance, individuality, and mutual respect.

This is just a fact of life—you see what you expect to see, what you fear you'll see, what you've been programmed to see, and what you hope to see. Therefore, to shift into a fresh quantum mindset, you'll first need to look at your mindset filters, those one-liner rules that you inherited from your family heritage and upbringing. Recognize the beliefs, attitudes, assumptions, and rules that were ingrained into your mind. From your own personal list of one-liners, choose which beliefs, attitudes, assumptions and rules you want to change in order to gain more freedom of choice and action in your life?

Before the fifteenth century, nearly everyone believed in the classic world view that the earth was flat, and that you'd fall off the edge of the world if you went too far in any direction from your home. Today this seems ludicrous and laughable, but back then it was a very real "fact" everyone believed to be true—if they thought about it at all. Then some crazy scientist named Galileo had the audacity to insist that the world is actually a giant spinning ball whirling around the sun and lost in infinite space, rather than an unmoving flat playing field at the very center of God's creation.

Most people back then, when told they were precariously straddling a celestial ball hurtling through empty space—that they were nothing more than a tiny speck in the vastness of an infinite universe—simply refused to believe the truth. Making the world view shift from Biblical to Newtonian reality was a massive leap for humankind's consciousness, and it took hundreds of years to accomplish.

Ever since then, science and technology have been continually expanding our personal sensory reach beyond our body's five external senses. We invented radios and televisions and can now see live-feed video images from our space station showing quite clearly what we can't see from the ground—that, indeed, this planet of ours is round and spinning and hurtling through the vastness of space.

This fact was not a readily-observable everyday fact, and it violated many deeply-ingrained beliefs and sensory perceptions. It was a fundamentally new idea that drastically changed the underlying way people thought. With this expansion in world view, humans were freed to venture forth and explore the planet, to imagine and invent new ways to go places faster, and to do things that our old "flat-earth mindset" couldn't even imagine. The whole world's mindset took a quantum leap into a new sense of life's potential.

The mindset shift we are approaching today is equally mind-blowing, if not more so. When we do something as seemingly simple and straight-forward as to look at our own hand, for instance, we're now being encouraged to see not just a solid curved surface of living flesh, but also a deeply mysterious phenomenon of invisible atoms made up of even more invisible protons and electrons.

We're also being informed that our brains are constantly busy generating a visual and tactile illusion of solidity, when in fact, almost all of what we think of as solid is just empty space with a few electrons and protons and neutrons maintaining a stable energetic structure.

The illusion of matter doesn't end there. We now know from experimental fact that even when we dig down right to the tiniest subatomic particles or quanta that make up all matter, there's yet another level of mystery that truly baffles the mind. It seems that fundamentally there's no such thing as matter at all. Instead, all quantum particles are made up of coded waves, of pure energy, and this universal ocean of undulating vibrating waves of energies have no definite material existence at all.

Instead, all we can count on in science at this point is the questionable assurance that at any given point in time, there's a probability field indicating that perhaps the wave will materialize as a particle at one place, or perhaps at another place. Nothing's for certain anymore. However, since there are many possibilities, it's our free will through conscious choice and decision making that shapes our destiny.

LIVING SPACE

It's no wonder that, seven or eight decades after Einstein first blew the top off the Newtonian world view, most of us are still having great difficulty seeing the world in the new quantum perspective. After all, we can't even see for ourselves what we're being told exists. Columbus and Magellan could at least sail away and find out for themselves if the world was flat or round. How can we dive down into nano quark-land and see what the physicists are talking about?

When we look in the other direction, up at the skies and stars and heavens above, it's no easier to grasp the quantum mindset. We're being told that there are literally billions of galaxies out there, each with billions of stars similar to our sun. Furthermore, there are millions of vast dark holes in our own galaxy, and all the empty space between stars is now considered an illusion because dark matter and dark energy fill up all that space.

There is an infinite number of intelligent nano-tiny bundles of energy throughout what we once thought was empty space. These energetic bundles have their own individual alertness and integrity, and function within tiny charged energy fields that connect each of them with everything else in the vast universe—and perhaps beyond.

The emerging quantum mindset challenges us to believe that this startling new vision of reality is more true than the old version. We're being offered the opportunity to shift our world view from feeling like we're isolated material bodies in a pre-determined static universe, to feeling intimately connected energetically not only with each other, but also with the infinite intelligence, compassion, and creative power of a sentient universe—yipes!

Furthermore, we're being encouraged to consider that we ourselves contain infinite universes within universes. A physics friend of mine was saying recently that he now believes that reality stretches off into infinitely smaller and smaller realms of being, and also infinitely larger realms of being, and that our personal consciousness is intimately involved in all of those infinite realms of being throughout many universes.

If this is all true, then the new quantum mindset we're considering leaping into is perhaps an infinite mindset. The human mind always wants to measure and quantify things. This is how we learn and try to make sense of the world. However, one cannot measure the infinite nor compare it to anything else that's not infinite.

Yes, we're biologically programmed and ego-bound to not drift off into that "infinity realm" too far—otherwise we might lose our earthly focus and die as biological entities on this material planet.

Our reflexive biological focus on local material reality doesn't mean that infinite realms don't exist, or that we're disconnected from them.

It just means that we must maintain our mortal conscious limits so we can sustain for a while longer this particular bundle of human energy that we call us. Meanwhile, we can reach with our quantum minds to touch the skies because the infinite mindset is the quantum mindset. This is the way God thinks and how we should think as well; infinite love, infinite time, infinite energy, and infinite possibilities surrounding us at all times.

Chapter 2:

THE FEELING OF REALITY

Notice I've just sent into your mind a number of provocative ideas, images, concepts, and possibilities—some that you've already encountered before, and some that perhaps you haven't. This is how your quantum mind grows, through opening to receive a new model or structure for reality, and then perhaps reflecting upon this model so it grows new synapses in your brain and generates an internal neural presence within you. We want to grow subconscious neural pathways that pave the road to what we want as if it already belongs to us.

If you just read quickly through this book and never again think about such new ideas and possibilities, then these ideas will die in your mind and fade off and away, and your mindset will remain caught in the old model. But if you continue to reflect and think and ruminate about a new possibility, you'll actively be growing your quantum mindset. Neuroscientists have found that over two million new sensory inputs rush into your brain every second, but you filter the vast majority out of your conscious experience so that you can focus on what you consider most important.

There's a part of your brain called the Reticular Activating System (RAS). It's a filter in your brain that teaches it what to notice while filtering everything out that's not important to you at that time.

For instance, think about the time you were car shopping and just bought a new car. Before buying the car, you rarely saw that model on the road, but when you were interested in buying it, suddenly it was everywhere. It's the same with certain shoes, purses, and watches you purchased in the past. The question is: *Was that car or item around before you started noticing it?* The answer is, of course it was, you just didn't notice it. But why not?

It wasn't until you purchased that item that your brain decided it was important. Once that was established, your RAS radar tuned into that item, magnified it within your reality, and attracted it into your life. As they say, where focus goes energy flows. This is true, because whatever you focus on or obsess about seems to attract itself to you.

When you set a goal or focus on a dream you're passionate about, your brain knows it's important to you and your RAS kicks in to help you. Your ears open and your eyes are now peeled. You start noticing yourself overhearing conversations related to what you want, new articles jump out at you, and people who can help you achieve this goal begin gravitating towards you. You begin to notice everything associated with this goal or focus.

This is why it's so important to set goals and to know what you want because your RAS is like a heat-seeking missile and will always find its target once it knows where it's going. The problem is that most people focus on their fears and what they *don't* want to happen, instead of what they do want. Their targets are set on the past and worst-case scenarios. Unfortunately, your RAS will work for the better or the worse, no matter what, because you're its commander-in-chief and it simply obeys whatever you focus on.

> *"The secret of change is to focus all your energy not on fighting the old, but on building the new."*
>
> —Dan Millman

If you (like most people) continually bombard your senses and your brain with stimulation from your TV, radio, cell phone, computer, and so forth, you deny yourself the reflective process that enables new concepts and mindset models to grow in your brain. You are the only one who can control the balance you maintain between receiving stimulation and reflecting upon that stimulation.

As a biological organism on planet Earth, you do need to focus regularly on sensory interactions with the world around you, so that you remain informed and respond appropriately in your everyday activities. But you also have the choice (and responsibility) to pause regularly to quiet your mind, calm your emotions, and look inward to tap into your insight and integration functions.

When you build some reflective time into each day, where you let the dust settle, calm your emotions, and reflect on what's most important, you open up to a higher level of seeing that integrates sensory and conceptual inputs into a meaningful sense of who you are and what you're doing. I'll provide you with several basic methods for doing this reflective process. What's important is to value this reflective process so that you take time to focus inward, get centered, and turn on this higher-order contemplative function of the mind.

As Professor Arnold Mindell explores in his remarkable 700-page book called *The Quantum Mind*, the basic insights and realizations emerging from quantum physics research were discovered several thousand years ago by employing the exact opposite method that scientists now use. Rather than looking outward at physical expressions of the laws of nature and the universe, ancient meditative masters all over the planet developed meditative techniques for focusing full attention inward.

What did they discover? Somehow, they intuitively discerned all the core principles of quantum physics. Without performing a single experiment, they accurately described the workings of the atom and the energetic base of material existence. Furthermore, they taught that beyond the dualistic dimensions of perception where opposites attract and generate pairs (like in the nucleus of the atom) there is a quality of consciousness and underlying reality that is unified, whole, complete, integrated, and most definitely conscious.

The early meditation teachers showed that by entering into this reflective quality of awareness, a meditating person can experience a higher state of consciousness that is fully merged and one with the Creator of the universe. This is of course a spiritual process and experience, not a scientific one, but the quantum realizations are the same. What the scientist is looking at from the outside, and the meditator from the inside, is in fact, the same core reality.

Scientists explore this universal reality by using mathematics which has been called the language of God. Its logic enables mathematicians like Einstein, Bohr, and Heisenberg to follow almost to the ends of the universe certain universal deductions—like those in Einstein's equation about the relationship between energy and mass stated as $E=MC^2$. Of course, math on its own doesn't tap into the non-logical dimensions of consciousness related to feelings, emotions, and the no-thought contemplative state of oneness with the universe. However, math *does* approach eternal wisdom when it includes the imaginary-number dimension of reality.

Similar to Einstein's and Pauli's seminal conversations with Carl Jung in the early 1900s, in 1965 through 1985 the free-ranging scientist David Bohm and the great independent spiritual teacher Krishnamurti discussed at length (in over 25 recorded meetings) how the spiritual and the scientific realms of inquiry have reached the same core conclusion, that the universe is infinitely complex and will always remain mostly a mystery to the scientific mind, but is readily available to be experienced through deep inward reflection.

We'll return to this theme in later chapters, with some logical ways to actually accomplish this inward-looking process on your own, to see for yourself what the quantum mindset is about. Here at the beginning of our discussion, I simply want to make clear that entering into the quantum mindset isn't anything new.

What is new is that through science, we've gained a more practical approach to discovering an expanded vision of what life is all about. We've come to hold science as a new universal religion that we can all believe in because we're convinced science is documented to be true. As we begin to let ourselves believe in the quantum model of reality, we can readily expand our quantum mindset and in turn, learn to tap a higher-order perspective and consciousness—without having to enter into serious lifelong meditative retreat. Combining quantum science and regular meditation (or contemplation) seems an optimum tech-spirit merger.

There's an adage in our culture, "Seeing is believing," but in fact, new neural research on our body's visual apparatus has shown that we actually "make up" most of what we "see" in our mind's eye. Even though we have millions of rod-and-cone receptor pods in the eye's retina, there are only about ten long neurons that link the retina to the optic center in the back of the brain. This means that the amount of visual data flowing into the brain is relatively small, because it's funneled through so few conduits in the optic nerve.

The result is clear: the visual cortex in the brain takes this raw external data, compares and contrasts and loops and filters the data, and mixes that information with memory banks of similar visual data from one's entire lifetime, and then finally (in around .5 sec) generates an inner visual experience that is mostly made up, like a dream or our imagination—not at all like a photograph or video. This means our visual experience is hardly a reliable source of "objective" information.

The experimental method has tried to overcome this perception-deficit by developing, over the last five hundred years or so, external perception-devices (like telescopes and microscopes) to look as deeply as possible into nature. But quantum research shows we've reached perhaps the limits of what we can see through any external examination and manipulation. The quantum vision that can be measured is nearly at its limit.

We are reaching the experimental point where the smallest quantum particle suddenly turns into something that is not solid, not material at all. Roughly 68% of the universe consists of dark energy, and dark matter makes up another 27%. All the rest—everything in our traditional material universe, everything observed with all of our fancy instruments—adds up to less than 5% of the universe. We simply cannot see or detect dark matter and dark energy; it's just too small and too unpredictable and too weird to observe with even the most sensitive technological eyes.

No one knows exactly what energy itself actually is and where it originates. Perhaps energy is as close as we can come to the Creator of everything. The good news is that we can directly feel this energy. In kundalini meditation for instance, we can purposefully tap into the energetic forcefield in and around our physical body, and also manage that energy to our advantage.

Beyond all math and science, each human being carries the ultimate research tool—consciousness itself. Beyond words, concepts, and our external senses, our own consciousness can choose to focus directly inward toward our individual source of consciousness, and experience intimately our non-dualistic oneness with the universe itself, and who knows, perhaps also with the creative power and presence that generated the universe into existence.

You most likely won't need to drop everything and head off to a monastery or ashram and meditate for years to tap into this level of the quantum mindset. But in order to access the inherent power of quantum consciousness, it's important to regularly pause and reflect. Follow the lead of many quantum scientists and experiment with basic meditative techniques to see if they suit you. If you do this, you'll discover for yourself what millions of human beings have already found—that your own individual mindset is capable of connecting with, and temporarily merging with, the seemingly-magical and eternally-mysterious unified realms of consciousness, which Bohr, Heisenberg and others often talked about.

A CONSCIOUS UNIVERSE

There's been a lot of speculative talk about quantum consciousness, and materialist-based physicists and science pundits have been ruthless in putting down such talk as nothing more than quasi-science. But even Werner Heisenberg insisted that, "The reality we can put into words is never reality itself," and, "The elementary particles themselves are not real; they form a world of potentialities or possibilities, rather than one of things or facts." He also said, "The first gulp from the glass of natural science will turn you into an atheist, but at the bottom of the glass God is waiting for you … scientific concepts cover only a very limited part of reality, and the other part that has not been understood is infinite."

If we had adequate eyes to see into its tiny depths, the universe would not appear as a vast empty space with occasional stars and black holes. The universe is chock-full of energetic wave bundles that sometimes manifest as quantum particles, each carrying its own coded information and impact potential, its special bundle of quantum information and its own unique interactive intelligence. This list doesn't even include the even-more-mysterious ingredients of dark matter and dark energy.

In other words, the universe is full of a mysterious substance that includes information, movement, and energy. Furthermore, all this information and energy is connected. We inevitably come to the same logical conclusion Bohr, Buddha, Einstein, and the others did, that the entire universe is one perhaps-infinite buzzing realm of consciousness.

Here's the punch line: you and I are not separate from this vast universal consciousness. As we'll discuss later, it's an illusion of the ego that assumes we're separate isolated bubbles of awareness. Each of us is in quantum truth, always intimately connected with and indeed one with everyone and everything in the universe, yet still having free will and individual choice over what we think and believe, which contributes to a greater collective.

This conclusion emerging from deep within the quantum-science community is that David Bohm's notion of the universe's "implicit wholeness" echoes Albert Einstein's conclusion, "Behind everything that can be experienced there is something that our minds cannot grasp, whose beauty and sublimity reaches us only indirectly ... everyone who is seriously involved in the pursuit of science becomes convinced that a Spirit is manifest in the laws of the Universe—a Spirit vastly superior to that of man."

These days the notion of Spirit isn't used often by researchers, but their new scientific models of the nature of the energetic interconnected universe enable us to sense in our quantum minds that our relationship with the whole universe can be experienced through our own inner sense of heartfelt oneness with something that's just as alive in Spirit as we are. Rather than looking up at the heavens or down at the earth and seeing nothing but inanimate matter and empty space, we can begin to grow in our minds a quantum awareness of our constant participation in a greater universal whole.

Your ongoing act of developing this inner ability of "seeing in a new way" is an integral part of leaping into your own quantum mindset. Believing in things unseen used to be a religious notion, but it's also a scientific practice, especially now as we may be reaching the limits of our perceptual exploration of the universe.

What I'm recommending is that you tentatively try on the belief that right now you're an integral part of a vastly larger whole— that even though you can't see physically into the infinite depths of life, you can open your inner experience to intuitively "feel and know" your relationship with the universe.

By entertaining this belief in what quantum science is laying out before you (and what mystics have seen from the inside-out for countless generations), you create an open space in your current world view for growth and expansion. You can choose to leap into a qualitatively new sense of who you are and learn how to use this expanded vision to manifest what you want in life.

As quantum scientists insist, at least for now the deeper mysteries of the universe can't be dissected, manipulated, measured, and observed, but they can be honored, consciously participated in, and sensed intuitively. That's where we're headed in this book and guided programs. When you courageously expand your beliefs to include what quantum research reveals about your deeper nature, you can gain access to your own full potential. Through this mindset expansion, you can consciously put to use the creation process through which we have all emerged.

So, I recommend that you honor the scientific method and what it's revealing to you equally honoring your inner experiences, your intuitive flashes, and meditative realizations. To again quote David Bohm: "There is something that our minds cannot grasp, whose beauty and sublimity reaches us only indirectly."

Here's a short experience to help you expand your world view further into the quantum mindset. You'll also find in-depth guidance on our accompanying website. If you take time to do these quantum meditations often, you'll accelerate your progress.

Exercise: "Seeing In A New Way"

Even while reading these words, at the same time begin to focus your attention inward toward the feeling of your own personal presence ... just expand your awareness so that you're paying full attention to your inhales and exhales effortlessly coming and going ... and also be aware of your heart beating in your chest ... your whole body is sitting comfortably as you continue breathing and reading ... be aware of this distinct energetic part of the universe that is you ...

Experience how you are the living breathing center of your own personal universe ... and feel how your personal universe is an integral part of the whole universe ... and beyond!

After you finish reading this paragraph, feel free to close your eyes for perhaps twelve relaxed breaths ... allow your awareness to effortlessly expand to include your breathing ... your heartbeat ... your whole body ... and let your awareness keep expanding as you tune into feeling your natural energetic connection with the universe ...

FEELING YOUR WAY

Often people assume their personal mindset is nothing more than a cognitive function of the physical brain, a set of underlying beliefs and assumptions, expectations and limitations that define their personality, their outlook, their dominant thoughts and attitudes, and thus, their behavior. In part this is true, and we're going to explore in later chapters how the subconscious "belief" function of the thinking mind tends to dominate our conscious thoughts and actions.

However, we are definitely much more than thoughts and beliefs. We're also continually immersed in our own sea of inner feelings—our full assortment of emotions, hunches, gut reactions, and intuitive realizations, not to mention our underlying feelings of inner balance, security, confidence, appetites, sexual passions and so forth.

These feelings can be stimulated by our five external senses and by our thoughts and future projections, imaginations, memories, impressions, and forebodings. And all of these cognitive functions can be stimulated by our inner feelings. They're all part of a remarkable feedback loop that generates our ongoing conscious experience and actions.

Our underlying feelings are also grounded in deep heart-centered levels of intuitive understanding and whole-body impulses and realizations that let us know what's most important, real, and true in our lives. This deep visceral level of our "feeling consciousness" is an integral part of our personal mindset.

During your shift into a quantum mindset, your feelings, thoughts, and beliefs will undergo a transformation. You'll discover that rather than remaining stuck in bothersome old emotional contractions and fear-based subconscious reactions, you can progressively let go of the negative beliefs and self-defeating one-liner rules that you inherited from your family and culture.

In this new mindset, you will be able to consciously focus your power of attention on whatever particular thoughts, expectations, beliefs, and rules that will free you to create a brighter future. This inner process will lead you into position to tap heightened creative power, as you step-by-step put aside anxious feelings and just say no to self-defeating beliefs.

As you transcend attitudes that breed depression and powerlessness, you'll become free to leap into positive feelings that make manifestation and transformation possible. The challenge is to integrate your thoughts, feelings, physical senses, and your deeper inner guidance. This expanded interplay of your whole conscious presence is what makes the quantum mindset so valuable.

THE FEELING OF CHANGE

Changing your mindset is just that—change. Even though we know change is ongoing and inevitable and often what we desire, we still often resist and avoid any change in our lives, because change can be scary as well as rewarding—and yes, sometimes change can be downright damaging and destructive. So right from the start, let's talk about the feeling of change, and how you can approach change proactively to ensure that the feeling will be positive and the results of high value.

Bruce Springsteen once said, "I usually feel uncomfortable before I go on stage to perform, I feel all this anxiety, my body's shaking—and that uncomfortable feeling is how I know I'm ready to go on stage." In other words, he observes his feelings even if they're uncomfortable and gives them positive meaning related to the challenge at hand. For him, going on stage is scary and always a risk, but he leaps forward anyway. He responds to the feelings in his body not with cowardice and avoidance of the situation, but with courage and the expectation of a positive outcome.

I encourage you to take the same approach to changing your mindset, your world view, and your position in life. Change is an adventure into the unknown, so of course you will feel both excited and a bit afraid to take the leap. A quantum leap is especially challenging because a quantum energetic shift usually doesn't happen as a gradual easy ascent into a higher plane of awareness. At some point, you'll be challenged to completely let go of the old to leap forward into quite a different perspective on life.

When you face a challenge that might change you, it's natural to feel shaky and weak and foggy—this is your gut reaction to change, telling you: "Hey, your life is changing—heads up!" Impending change, especially of the interior type, will almost always bring you face to face with the quantum unknown where multiple probabilities offer a chance of something quite new happening in your life.

At the moment of choosing to leap, you encounter the core of quantum physics. You become aware that at any given instant, there are always multiple choices facing you, and your life will change based on the choices you make. Of course you feel a bit uncertain in the face of change. That's the name of the game: take the choice, make the leap, and transform your life.

Based on Heisenberg's uncertainty principle, you can never really know beforehand what's going to happen. Ultimately, you're forced to surrender to the laws of probability, and all you can do is accept what you discover on the other side of your leap … and carry on. But by using your thoughts and focused energy to a directed goal, you can move the probability closer to the desired outcome.

The lesson here is that even uncomfortable anxious feelings can be of great value. They alert you when you're on the edge of change, and they wake you up so you're fully "present" as you encounter the looming change. Your feelings serve you by emotionally noticing the approach of change, then charging you so that you can respond to the change appropriately.

When Bruce Springsteen feels anxious and uncomfortable when it's time to step onto the stage and perform, he uses this anxious reaction to the situation to push him into a high charge of energy. He taps the emotion of courage to propel him successfully into action. He believes these energies and emotions are what make his shows great. You can do the same when you feel uncomfortable in the face of change—use that emotional charge to propel you forward into action rather than backward into retreat.

You've probably noticed that many people habitually block their uncomfortable charged feeling in the face of change. You've seen the result—they don't experience any growth or breakthrough. They stay in repetitive low-risk comfort zones where nothing is ventured, and nothing is gained.

The body is naturally conservative; it reflexively avoids risk and pain, but that means it also avoids possible transformation. You can continue to stay in your comfort zone where not much is happening nor changing. It's safe, but it's also boring and limiting. Insights from quantum physics will help you be brave, take risks, and venture into the unknown.

HONOR YOUR FIVE SENSES

Your natural ability to be sensitive, to detect and interact with the world around you through your five external senses, is what enables you to survive and thrive on planet earth. By definition, all living organisms are sensitive and responsive; there's a sensory level of consciousness even at the cellular and microbial level of life. All the tiny cells in the body can receive stimulation from the outside world, identify and process that incoming data, and respond or react accordingly.

This is the core power of a sensory system. As an individual unit of life, it can sense and interact with the multitude of other sentient beings all around it. This is what makes the whole thing of life work—we're both distinct separate individuals and we're also inter-connected through our senses with each other so we can work together. The more advanced and complex life becomes, the more our senses have evolved in response to that complexity.

Many other animals have more sensitive senses than humans. Just think of a hummingbird's eyesight, a dog's hearing or sense of smell, a bee's sense of taste and location. We are comparatively weak in those senses, but luckily, we have brains that can do many unique things.

It's important to note what we've done recently to expand and extend our sense of connection. The internet and cellphone revolution has been one great leap into a higher dimension of sensory attunement. Now we can instantly see a very clear image of our loved ones when they're on vacation a thousand or ten thousand miles apart from our physical senses. We can listen to a complex musical concert recorded twenty or fifty years ago and hear it as if we were right there in the concert hall. Our mastery of electricity has empowered a vast rapid expansion of our sensory experience—with who knows what ultimate consequences.

Ancient mystics in all cultures have insisted that our senses connect us with our creator, and that we are the eyes and ears of God. From the quantum perspective, this seems to be scientifically true. Our senses record and process what's happening in the outside world and within us. In a conscious quantum universe, this information becomes instantly available everywhere. This seems to be what the ancient Hindu masters and the Theosophists were talking about with their idea of the Akashic Records.

Our senses are stimulated primarily by invisible waves—light and especially sound. At more subtle levels, our brain continually broadcasts information outward energetically (as in EEG research that picks up the electrical off-gassing). Furthermore, our whole body is an electromagnetic force field that extends outward into the environment, interacting with the force fields of other bodies'.

So, when we talk about our feelings and our senses, we're talking about ourselves and others as immensely complex resonance fields. Our inner feelings are not just our own; they are continually impacting other people, and we're also being continually impacted by others. Whether we like it or not, we're all living in a shared-resonance field. Scientifically, very little is known about this yet, but we're all able to sense the situation. With a quantum world view, we can all begin to take more responsibility for how our feelings are touching others.

AS ABOVE, SO BELOW

When we look at how our feelings fit into a quantum mindset, Newtonian science can explain the neuro-chemical functions and glandular secretions that stimulate or depress an emotion. Psychologists can explain how our emotions helped us to survive back in the wild. Now with new quantum insights into our feelings, we can begin to explore the possibility that our human feelings, especially positive unifying ones like compassion, empathy, trust, enthusiasm, and so forth, are somehow integrated into higher-order trans-human resonances and responses.

Many people believe there is a God up in the heavens who was very much like human beings. The Bible even declared that we were formed in God's image; the Old Testament portrayed God as having all the emotions that human beings do, including anger, compassion, jealousy, forgiveness, impatience, and hope. The Christian revolution centered around belief in a God who felt so much love for human beings that he sacrificed his only begotten son to save humans from their hopeless condition as sinners. Other ancient religious traditions such as Hindu, Taoist, Islamic, and Native American, also gave their Gods the same emotions as humans experience.

Then came the scientific revolution that judged such notions as seriously anthropomorphic, projecting onto an infinite distant Godhead or Creator our base animalist human qualities that had, according to Darwin, evolved out of a particular biological need, not a celestial dictum. The Newtonian world view scoffed at the very idea that there was some invisible God up there who looked like us and had come down and inseminated a poor Jewish girl in order to create a half-God, half-human savior.

But now, brilliant academic scientists are considering another and perhaps higher and more realistic logic when they think about human beings in relation to the cosmos. Professor Mindell comes right out and declares that there must be a direct link between our thoughts and emotions and the universe as a whole. Why? Because logically, from the very beginning of the universe, everything in the cosmos has emerged from one unified source.

This is what David Bohm seemed to be referring to when he talked about the implicate order: the whole universe was created from the same stuff, had the same origin, with the same template and matrix and so forth. From the one emerged the many. The universe might seem infinitely complex, but quantum science has shown us that the building blocks of the entire universe are very few. In fact, 93% of our body mass is made of stardust.

By logical deduction, this means that you and I cannot be separated from the Creator of all this physical and energetic reality, no matter what our theological, philosophical, or scientific perspective might be. Deductively, everything in the universe was created in "God's image."

That brings us back to our personal feelings, our emotions, and our inner experience of being alive and responding to the world around us.

Whether we like it or not, the quantum view is that all of us, including our thoughts and our feelings, are deeply rooted in a common creative source. This is the foundation of the quantum mindset perspective.

I mention all this not for any religious purpose. This book has no theological preference. I want you to seriously consider the possibility that your individual thoughts and feelings aren't only of evolutionary origin related to survival of our species on this planet. From the quantum view of astrophysics, it's within the realm of probability that you are in essence a highly-related chip off the old cosmic block. The emotions that you feel might be of a much more universal quality than Newtonian science ever imagined.

I reiterate this because for hundreds of years, science has been denigrating our inner feelings, insisting rather rudely that reason was king, and emotions were obsolete. Researchers fixated on quantitative data to the exclusion of qualitative experience (thoughts, feelings, imaginations, visions, and insights). The classic assumption has been that the experimental method would someday explain away all our emotions and realizations in purely materialist terms.

Quantum science discovered that our thoughts and emotions are functioning not just as biochemical reactions to be analyzed but as concrete phenomena in the space-time continuum. Our thoughts and intuitive feelings are energetic phenomena, and very possibly occurring outside the space-time continuum, as we'll explore later. Furthermore, there's a high probability that our inner feelings, as energetic wave flows, radiate outward and have an impact on all the wave flows around us, as the next chapter will highlight.

When it comes to manifesting anything on the physical plane, all our abstract ideas and plans and cognitive activity don't accomplish anything at all, unless we include in the manifestation equation the energy fields of our driving emotions: motivation, personal power, hope aspiration, courage, and all the rest. Without our feelings, nothing at all would get done. Take away our emotions and we're powerless—this is important to remember!

Human beings are in reality very similar to the old religious notions of the Creator, in that we can personally muster the power to create what we want, by using our imagination, our passion, our instincts, our intelligence, and our bodies to create our future.

Originating in your inner non-material feelings, dreams, and passions, something quite material in the outer world can come into being. If you want to learn how to manifest what you need and desire at higher levels, it is vital to honor and understand, integrate and apply all aspects of your creative potential. Your quantum mindset will of necessity include your thoughts and your emotions, your ideas and your feelings, your plans and your passions.

You do have the original spark of your creator in your physical, mental, and emotional body. As you expand and evolve your quantum mindset, you will need to focus on, develop, and transform all three of these manifestation elements.

Chapter 3:

QUANTUM RELATING

How you relate to yourself and those around you is determined by the basic laws of quantum physics that we're exploring. The entire universe is governed by universal forces of attraction and repulsion, as we'll explore in this chapter and the next. These forces can be applied to your personal and professional relationships to maximize bonding, trust, creativity, and cooperation.

In this chapter, we will do a deep dive into the actual science of quantum mechanics, so you understand the principles of how subatomic particles attract and repel each other. By extension, you will learn how to move effortlessly away from relationships that don't serve you and attract people into your life who are naturally drawn to you.

Four thousand years ago in the Hindu Vedas, spiritual teachers wrote of a primary God called Shiva who was both the destroyer and creator. The Judeo tradition saw Yahweh as the same, creating and destroying through judgment and passion, anger, and love. We live in a quantum universe full of creation and destruction and new creation. Consider the physics of a black hole to understand the magnitude of this creation-destruction-creation power.

Naturally the question arises: What particular feelings and emotional charges should you nurture to become a creator and manifest your own inner visions? For instance, sometimes you must destroy what exists so you can create the new. Often you must end something to embrace a fresh expression, so even the supposedly negative emotions that propel you into destruction should not be totally discounted.

To create something, Jesus and Buddha had it right at quantum levels. God is equated with one particular emotion: God is love. Empedocles in 450 BC declared that love is an attractive power in the world. Also common was the idea that compassion generates creation. You probably already know in your own life that the spirit of loving and tolerant inclusion will empower your higher levels of discovery and expression. If you don't feel love for what you want to create in your life, chances are you'll fail at your endeavor.

So—what's the general law of creation? If your emerging quantum mindset is grounded in *love*, you will maximize your success rate.

Creativity = Love x Vision

Creativity seems to require both material and non-material elements. It should be obvious that developing a quantum mindset will require you to let go of outdated Newtonian assumptions about your reality being static, predictable, solid, and material. Certainly, the Newtonian world view works for most larger endeavors on this planet, but when it comes to building the foundation of a quantum world view, the non-physical invisible "i" aspect of love *must* be included as primary.

Love is considered eternal, while at the same time it's always changing. And if love is the foundation of your new world view, then your mindset is continually evolving, emerging, and expanding into unexpected realms. You enter a playing field where your limited sensory inputs and conditioned attitudes and assumptions interact dynamically with, well, the whole universe, and it's all happening right now! Your world view is continually in process. It's responding to new insights and imaginations that are often inspired by your personal consciousness engaging with a greater consciousness.

Quantum physics leads us to accept the higher reality that the whole universe is indeed one unified conscious presence, and at deeper energetic levels we're all part of that one seemingly-infinite mind, just as the ancient masters suggested. We receive most of our best ideas and flashes of inspiration right out of the blue, don't we? Such intuitive realizations are most likely a subtle interaction between our personal limited consciousness and the full integrative wisdom and power of the universe—and, who knows, perhaps beyond.

In this spirit, quantum physics brings the mystery back into science—it's created a world view in which all levels of human experience can be included. Great scientists like Niels Bohr and Arnold Mindell have documented how the dreaming-states and visions of traditional shamanic masters are completely 100% congruent with the imaginary-number, dark matter, anti-particle realms of physics and mathematics.

Our own lives express this. We live in an experiential world where our dreams, while we're dreaming them, are just as real to us as our waking reality. Our subjective insights can seem just as valid as our objective deductions, and our imaginations just as valuable and powerful as our physical expressions. The current media interest in the phenomenon of lucid dreaming reflects this growing realization that dreams are somehow real.

As a modern society, we've done our best to pretend we're no longer superstitious or religious, that we've shed our mythic-imaginary tradition and entered into a pure understanding of what's scientifically real. Then quantum physics comes along with its insistence that all the deeper mysteries remain alive and well—both out in the universe and inside our own heads and hearts.

Probability has replaced certainty, and participatory freedom of choice has replaced fated determinism. We're plunged again into a world view in which we can't pretend to control everything, but we've gained a new freedom to consider alternatives, to experience oneness, and to see ourselves as truly cooperative creative beings.

THOUGHTS BEYOND SPACE AND TIME

As we've seen, your mindset is a function of your physical brain, but research in neuropsychology still can't quite locate or quantitatively study the mysterious phenomenon called consciousness. Quantum physics is, however, providing new models to describe our thoughts and imaginations, our feelings and intuitions, in a truly meaningful way.

In Newtonian-based science, the brain was considered nothing more than a complex material phenomenon. Anything beyond the workings of the physical brain was put aside as fanciful religious belief, not scientific fact. With the advent of the quantum mindset, the brain itself has expanded into a new model. Specifically, the mind (as opposed to the brain) is caught up in the new quantum vision of reality where energetic waves rather than physical particles run the consciousness show.

Thoughts and imaginations, dreams and emotions leave traces of evidence indicating their existence. However, our inner experience, our awareness itself, simply refuses to be detected with scientific measurements. Traditional science was based on the premise that we can separate the observer from the observed but ever since Einstein declared that this objective-observer stance is a scientific illusion, we've had to try and re-envision what it means to be a scientist.

The emergence of what we call the quantum mindset reflects this traumatic loss of the Newtonian world view, and a shake-up at the very core of science. The dilemma is being at least temporarily resolved by accepting the Newtonian world view as valid in most general cases like building a skyscraper or extracting a tumor. But for the very large and the very small aspects of reality, including the creation of the universe and the subsequent (and somehow linked) emergence of human consciousness on our planet, we must call on a scientific Shiva to destroy our old world view, so we can create the new.

Let's say you come up with a great idea. It could be in a relationship, your work, your hobby, or your philosophy. You dare to ask yourself, *Where did this great vision, this new creative idea, come from?* Early Greek philosophers struggled with this question for centuries, as did the ancient Taoist thinkers in China. Even way back then they realized, using direct inner observation of the workings of their own minds, that thoughts, imaginations, dreams, and visions emerge from some creative source that seems to transcend our individual brain function.

> *Try this for a moment: imagine something, anything at all— perhaps a red rose, a fancy car, or someone you love. Hold the image that appears in your mind ... breathe ... and try to find where that image is located. There—that's the quantum question ... does that image you hold in your imagination really exist on the material plane? Is it real in terms of Newtonian physics? Can it be detected by experimental dials and measurements?*

In fact—no. When you conjure up an image in your mind, you're already dabbling in the larger life mystery that classic science tried to obliterate and failed to do so. The most astute researchers in consciousness studies like David Chalmers say, "Consciousness poses the most baffling problems in the science of the mind. There is nothing we know more intimately than conscious experience, but there is nothing that is harder to explain. Neuroscience alone isn't enough to explain consciousness."

If we take a leap beyond the confines of neuroscience into quantum mechanics, what do we find? We find, for instance, that our dreams don't happen within the spacetime continuum that Einstein described. Can you remember a recent dream and the images you experienced in that dream? Did they take place in the present, in the past, or in the future? As Carl Jung pointed out a hundred years ago, they occurred in none of those times.

Dreams are ephemeral; they seem to happen in an alternate flow of time. Even the space in dreams is qualitatively different from space in our waking experience. It's the same with the rose or person or whatever you imagined in the above exercise. Where does that image exist? It exists in your mind, but it doesn't have any material substance at all, certainly none that science can yet detect.

But is that image just an illusion, a pure figment of your imagination? No—at some level you know it's real to you, therefore, it exists at some level. It impacted you, it stayed in your mind, you're still aware of it, and you're conscious of it! If you believe the image is unreal, then you'll relate to it as unreal. As you shift into a quantum perspective, all probabilities are possibilities, and you can trust your inner experience as equally as you trust your outer senses. Then, by honoring your imagination, you enter into a quality of consciousness that is inclusive and breaks free from materialist limitations. You're even able to play with that image, perhaps even materialize something that emerges from the initial dream image.

In practice, this is how anything new comes into being. We play in our minds with our dreams, our memories, our imaginations, our passions, and our challenges—and out of thin air, so to speak, up pops a fresh idea, a new plan, and a vision of something we want to act on and materialize. What was real first in our minds step-by-step emerges in full material glory.

This is how we bring newness into our lives. This is how we stimulate positive change. This is how we manifest anything at all, be it a new friend, a new house, a new job, or a new belief. It all begins at the ephemeral level of reality where we act as creators, and then, through the power of our will and intent, intelligence and diligence, we generate the result we have visualized. We use our minds to turn the qualitative into the quantitative, and we always begin with the imaginative process.

TAPPING QUANTUM POWER

Consider in your emerging quantum mindset the lurking possibility and tentative belief that your psychic mind is more than just your physical brain. If you believe you're nothing more than a predictable bio-chemical machine or complex bio-robot, then that belief will limit your inner experience. When you take hints from the new science, where all things material is not material at all, but tiny bundles of intelligent responsive energy, then your new belief will free you to consider your mind in a vastly larger light.

Specifically, if you're made up of intelligent wave forms in motion, rather than static atoms located in one particular place, and if the universe as a whole is one vast super-mind from which you have emerged and remain connected—then the possibility naturally arises that all your personal thoughts and feelings, images and intentions aren't isolated phenomena of an individual brain. Instead they might very well be interactive happenings inspired by a consciousness beyond your individual brain. This is a gigantic idea, a looming quantum possibility—and well worth contemplating.

From all traditional cultures, we have learned that dreams matter, that the subjective inner experience is important and often leads to more objective external expressions. All religions of the world came to a consensus realization that the universe was created by some infinite source or presence, and that our individual minds are somehow connected, perhaps even one with, that infinite creative source.

If we put aside theological dogma and think about this from the quantum science perspective, it seems logical that if everything in the universe was created by a larger force, a conscious energetic presence beyond the known universe, and if each of us is actually an integral conscious part of that creation, then the thoughts, imaginations, and feelings we generate are somehow also an integral outflow from that universal mind, by whatever name we might call it.

Another great physicist, Stephen Hawking, once said, "The whole history of science has been the gradual realization that events do not happen in an arbitrary manner, but that they reflect a certain underlying order, which may or may not be divinely inspired." Perhaps when we feel inspired, we are tapping into this mysterious realm where we transcend our brain's physical limits, and flash with a transcendent thought or feeling that takes us "beyond ourselves."

Quantum physics is now telling us that we're not deluding ourselves when we feel our new ideas and visions come from "beyond us." The feeling is subjective but that doesn't mean it isn't real. In fact, the quantum mindset insists that objectivity is the illusion.

As individual embodied consciousness, we can never really step outside our state of total immersion in reality. But our understanding of reality is now evolving. We must accept that we're made up of bundles of wave energy that radiate infinitely outward in all directions beyond our physical bodies. Our energetic wave-radiations interact with everything around us.

Here's the point: when we take the quantum leap into a new mindset and expand our beliefs to include higher possibilities than we imagined before, then we tap into the creative power not just of our individual brains but of the whole universe. We're continually connected energetically, if not bio-chemically.

Taking the leap into this expanded belief is what the quantum mindset is all about. This new vision shouldn't be considered quasi-science. It's best labelled as exploratory science, expanding our model of the mind so we can see ourselves functioning within an energetic rather than a material reality.

Do you believe that this new energetic model makes more sense than the material view? How does this discussion make you feel? Is your breathing tight and defensive, or relaxed and expansive?

I'm hoping in this book to show you alternate ways of perceiving your own mind, your own potential—and again, I encourage you never to take my word for any of this. Look to your own inner experience as well as the evidence of new quantum research. What rings true for you?

When you consider both the science at hand and your intuitive hunches, your gut feeling, and higher instincts, your whole being can participate in this exploration of your full potential. There's great value in taking your subjective inner feelings seriously. For instance, how do you feel about the notion of intelligent energetic waves interacting with each other through what's called resonance—where your personal resonance field projects outward and interacts with other people's resonance fields?

Ultimately, feeling is your core tool for relating and learning, for discovering and manifesting. Your feelings can guide you, inspire you, motivate you, protect you, and reward you. In the quantum mindset (which always focuses on wholeness, integration, and participation), all of your senses, internal and external, are a unified sensing system interacting together to generate feelings and thoughts that lead you forward.

To highlight this as an experience, let's turn now to a beginning process for integrating the various energetic and emotional centers up and down your spine. At any given moment, if you pay attention, you'll find that you're always having feelings and energetic flows of one kind or another moving through your bodily awareness. Using a traditional feeling/energy model for this, I'll help you focus on the actual experience happening inside you—the feelings that energize you to move into productive and enjoyable action, or to kick back and enjoy the moment. We'll also explore how to experience your feelings as an integrated whole. This is a 3-5-minute experience, which you can also find as an audio-guided program on www.TheQuantumMindset.com.

Exercise Two: Feeling In A New Way

Go ahead and get comfortable ... now tune into your breathing. Feel free to stretch a bit if you want ... yawn if that feels good ... and let go of any tensions you find inside you. Even as you read these words, first pause and notice what feelings you might discover flowing through your body ... where do you feel a charge of energy, and where do you feel blocked or low on energy in your body? Breathe into whatever feelings you find ....

Now begin to turn your mind's focus of attention down into your feet ... into your legs ... and up into your pelvis region. As you stay aware of your breathing, see if you feel light and charged and good in your lower torso, or if you feel heavy, out of energy, and dull in this lower region. See what dominant feeling or emotion you find down low in your body ... don't judge what you find, just observe—and allow the feelings to change freely ...

Now expand your awareness to include what's called the second chakra, the sexual and creative energetic center of your body, right where you'd expect to find it below your belly region. Again, breathe into the energy and emotions you find there ... and let the two lower energy centers come together as a unified pair ...

Next expand your awareness up into the third power center of your body in your belly and diaphragm region ... notice if you feel powerful in your belly and your breathing, or weak and low on energy. What feelings do you find in this region? Allow all three lower energy centers to feel integrated ...

You can now expand your awareness another step to include your fourth energy center, in and around your heart in the center of your chest. Breathe into whatever feelings you find there ... perhaps say the word "love"" to yourself ... and see what feelings are flowing through you in this region. Are you feeling strong and bright and expansive in this key area, or feeling weak and contracted and blocked?

Now expand your awareness to include your throat, tongue, jaw, and lips—the energetic communication center of your body. As the air flows in and out through your throat and vocal cords, notice if you feel tight or relaxed, open or closed down, anxious or relaxed? Breathe through your mouth if you feel under pressure ...

Now expand your awareness upward to include your mind and eyes and cerebral region in your head. Breathe and experience the volume and space inside your skull ... and perhaps say the word "insight"—and notice how you feel in your mind ...

Finally, expand upward to the top of your head, to your crown—where your personal consciousness meets whatever expanded consciousness there is in the universe ... breathe into whatever feelings and energy flows you might find here ... and perhaps say to yourself, "connecting" ...

Now allow your awareness to expand to include your whole body at once, here in this present moment ... feel yourself as an integrated energy system ... and breathe into whatever feelings and insights might come to you now ...

As you end this whole-body feeling experience, remember you can return and move through this exercise any time you want to charge and balance your feelings and energy. This process is ideal if you want to muster your positive feelings and energy toward envisioning and manifesting whatever you want to attract into your life. We'll be returning to this later in the book as well.

CONNECTING ENERGETICALLY

When we look very closely at life then pull way back for a galactic perspective, the primary quality of the universe is that everything seems related to everything else. We now know that each and every atom in the universe is held together by powerful relational forces that bond electrons, neutrons, and protons. At planetary and galactic levels of relating, gravity holds everything together in perfect order.

Let's look more deeply into the basic relational laws of quantum physics and continue exploring how these subatomic bonding forces might also influence and define how we relate with each other, and how we attract or repel people, things, situations, and synchronistic flows in our everyday lives. Feel free to move quickly through this discussion if you're already on top of the new quantum discoveries.

In the old Newtonian world view, everything was seen as relatively static; you could pinpoint and freeze-frame anything and study it objectively at your leisure. Likewise, with traditional psychology, an individual's personality was considered relatively static and rela-tionships mostly predictable. Before Einstein, you could get away with thinking of both yourself and the people you related with as pre-programmed entities with mostly predictable behavior.

But that's all changed now that we know that everything in the universe (including ourselves) is in motion, in process, and changing continually. Like our ever-expanding universe, we're also personally in a state of ongoing change, flux, possibility, and transformation, as are all the people in our lives that we relate with. We might continue to hold onto static concepts of who we are and who our family and friends and colleagues are, but we need to accept that those are just concepts we've developed, not the deeper reality of who we are and who we hope to become.

Quantum physics is transforming psychology, because the basic defining premises regarding who we are as individuals and how we're changing over time have forever shifted. Now that we know that everything in the universe is actually energy in motion rather than solid static matter, we need to actively evolve our world view to see ourselves and each other in the new mysterious light of probability and the uncertainty principle.

As we examine the experimental evidence of how quantum physics sees the core nature of relationships, we'll discover that each of us is a walking, talking bio-miracle created by almost infinitely-tiny interactions of intelligent, conscious and coordinated energy packets. We're not isolated distinct entities; we're continually dancing an integrated energy dance with all the other people around us, and by extension, with the whole universe.

We're also going to explore the quantum process called "entanglement" and show that it can be expanded into a new model of who we are as a whole organic being. For instance, "Quantum particles can interact in such a profound way that they lose their individual identity and behave as one," a recent article on quantum personality stated. "Moreover, the interaction results in a new entity with properties different from either of its constituents." This is a remarkable new model of "who we are" as individuals, and also predicts how we often get entangled in relationships and communities."

This entanglement dynamic might be a lot to integrate into your new world view, but great are the rewards of leaping into this new understanding of who you are and how you relate to others. The new model of relating frees you to experience your interactions with others in a much more intimate and creative way.

Mystics, lovers, meditators, and nature lovers have always sensed subjectively that we are all somehow an integral part of a greater invisible unified whole. Until recently, scientists couldn't verify the assertion that there's an inherent pervasive resonant level of relating that holds us all together.

Math and quantum physics lend credence to what wise folks have known, through inner intuitive experience, ever since the dawn of civilization and probably long before that. The ephemeral laws of attraction that seem to magically draw people together into lasting bonds and relationships are becoming more clear. As scientists identify actual physical processes at subatomic levels, we can expand our mindset to consider these processes as part of the core dynamics of quantum relating.

This is really exciting stuff. We all know the feeling of loving attraction to another person, but we never before knew that attractive love might be a power, or force that transcends interior feelings of romance and passion. In fact, love might be an expression on a whole-body scale of the very tiny processes that hold all our atoms and molecules and cells together in a larger integrated living organism.

As we delve into new insights from the still-emerging science of subatomic physics, regularly pause and reflect on how this subatomic universe deep inside every atom in your body and your friends' bodies reflects how you relate heart to heart, mind to mind, and body to body.

WE ARE ALL INTIMATELY RELATED

"You and I, while we sit and talk, do not feel quantum. We seem to have distinct outlines and do not crash and combine with each other like waves in a pond. The question is, why does the world look so normal when quantum mechanics is so weird?"
—Markus Arndt, physicist at the University of Vienna in Austria

Science now shows that everything in our universe is in relationship with everything else. Quantum research has been exploring this relationship dynamic for decades and has steadily expanded our prevailing mindset regarding the core laws determining how things relate. If we understand these universal laws and natural behaviors, we can also learn to relate more effectively with each other.

For instance, our universe was probably created (or reborn) from a zero point as most scientists believe. Right at that instant of creation or celestial rebirth, all of the energy and waves and particles in the cosmos were suddenly released all together. They came from a common source. From the very beginning, everything in the universe has been connected!

Over vast amounts of time, the original quantum particles in the universe spread in all directions. They also engaged with each other, forming the hundred or so core elements and then molecules and becoming complex bundles of intelligent energy/mass. A primary aspect of the quantum mindset is accepting and believing that right now, our entire universe still consists of that same original mass and energy, somehow still caught up in an ever-expanding universal matrix of gravity, motion, energy, structure, resonance, and entanglement.

What does this universal situation tell us about our everyday lives? Fundamentally, it means that even though we tend to see ourselves as separate isolated individuals, deep down the opposite appears to be true. You and I and everyone else are right now and forever energetically connected.

Everything is connected to everything else and that is the world in which we live. Consider this excerpt from the book, *THE WORLD IS YOURS*, by Kurtis Lee Thomas:

> No matter what religious beliefs one has, we must come to the realization that ultimately, we all come from the same place and we all have a spark of the life force from the creator within us. Take the example of our Prime Creator (Source, God, or whatever you choose to call this energy) being a giant body of water—an ocean in the sky let's say. Now let's consider humans as just a tiny drop of water from that ocean—even if we are just a single drop of water taken from that giant ocean of life, we are still comprised of all the same things that make up this ocean. And if we were to examine this single drop of water under a microscope, we would see that it is made up of the exact same elements of the larger body of water that make up this massive ocean.
>
> Humanity is just like these drops of water in the ocean. But if we were to ask ourselves "When the drop hits the ocean, where does this drop end and the ocean begin?" Hmmm. Well it doesn't, the drop simply becomes the ocean. Human beings are just small drops in our universe; which would therefore make us become one with the universe (the same as the drop became one with the ocean). We are merely a smaller replication/hologram of the massive universe we live in.

At subatomic levels this means that what you do affects me at very subtle (and sometimes not so subtle) levels. Furthermore, what I think, the emotions and ideas, and the dreams and fears I hold in my mind and body, will at energetic levels radiate outward and impact your thoughts and feelings.

> *"Nothing in Nature lives for itself. Rivers don't drink their own water. Trees don't eat their own fruits. The Sun doesn't shine for itself. A flowers fragrance is not to smell good for itself. Living for each other is Nature's rule."*
>
> —*Unknown*

Hopefully, as we all come to realize the quantum truth that everything affects everything, we'll finally learn to share, to be a bit more compassionate, and take more responsibility for how our personal lives are continually impacting the rest of the world.

Likewise, looking microscopically into our own bodies, it's equally true that each of us lives distinctly within our own amazing interior universe. Our organism has its biological capacity for integrated self-management. Somehow our billions of cells and microbes and so forth, manage to function in an almost-perfect state of harmony. Each cell knows how to participate in the greater whole of our body, maintaining balance, equilibrium, integrity and health.

We also possess a powerful, self-centered, conscious ego, plus a subconscious matrix that balances all our past experiences and assumptions, but most of the billions of cells in our body are running on automatic, functioning as an infinitely-complex universe within us. This mostly-unconscious system will continue until our inevitable mortal demise.

Along with our core individuality, we also have the natural need and tendency to do what the original atoms did. We attract each other, form dyad bonds, and multiply as families. Whenever we want to accomplish something in the world, we gather together as groups, communities, political parties, and religious movements. When we share a unity of purpose, we can accomplish great things.

This propensity to seek and form new relationships in our lives is the underlying power of quantum consciousness. Take it away and there is no human society. Take away this attraction tendency at *subatomic* levels and there is no universe at all. Our inherent yearning and ongoing ability to connect energetically, emotionally, physically, and spiritually is what empowers life itself. This natural human "tendency toward relating" is a vital aspect of survival, and a reflection of the fundamental energetic-bonding nature of the universe itself.

You might want to pause for a few breaths after reading this paragraph, look up from reading, tune into your breathing and the heart-felt feelings inside you right now ... and see if you can feel the experience of being connected with someone you love, heart to heart, right now. You are after all, a walking, talking electromagnetic forcefield that radiates outward ... see if you can experience this feeling of being connected energetically body to body with another person ...

Chapter 4:

QUARKS AND GLUONS

We're born into this world without clothes; we come in naked and possessing nothing at all. While we're children our needs are mostly taken care of by others who have manifested everything we need to survive. As we move into adulthood, we face the universal challenge of manifesting into our lives all the things and people and situations we need in order to survive and hopefully thrive.

We often think of manifesting as a process we move through to get the things we want—a new job, better car, or great tech gadget. This is all related to manifesting money, of course, and that's important. What's more important in our lives, manifestation-wise, is attracting special situations and especially special people to us.

The key word we've been discussing throughout this book is still front and center: the power and dynamics of attraction. Thus far, I haven't gone into the nuts and bolts of how quantum science understands the universal forces that bring things together into durable relationship bonds. In this chapter, I present a concise overview of the latest revelations about the universal laws of attraction. Then we can apply these laws to your personal situation and the relationships and situations (and sure, the things) you want to manifest into your life.

All of us who made it through high-school science classes were taught a general myth about how particles and atoms and molecules attract each other to form bonded relationships. Top physicists often point out that much of what is taught in school is out of date, misleading, and couched in vague concepts that fail to convey the reality beyond the concepts.

To update you on new insights in quantum physics that shed light on how particles and people attract each other and unite, let me share my notes on what's happening at subatomic levels of relating. You can then match my "science story" with yours and expand your mindset where needed regarding the universal matrix in which you live and relate. If you already understand this theory, or don't want to dive further into it, feel free to move quickly through the remaining pages of this chapter.

Quantum researchers, just like the Newtonian and Einsteinian pioneers before them, must face the challenge of expressing in everyday words what they're discovering in their miniscule experiments. In this communication effort, scientists generate contemporary science myths and models to hopefully enable everyday people to at least somewhat grasp what the new science is exploring and discovering. We need to remember that science, by definition, is based on the best models to date, not on any absolute truth.

That said, here's what seems to be the most verifiable and trustworthy view of quantum attraction to date: Most of us know from traditional physics that the building blocks of all matter consist of tiny invisible atoms with varying numbers of protons and electrons in their nucleus. Way back in 300 BC, the Greek philosopher Democritus described with remarkable accuracy the existence of what he called the 'atamos' (which means 'indivisible' in classic Greek). A short time later, Aristotle called the very idea pure nonsense.

Even further back in history, using deep-meditation tools for exploration, great thinkers in China, India, and elsewhere sensed the presence of tiny building blocks deep within solid matter. Heisenberg, Bohr, and Schrodinger read The Vedas and Upanishads, for instance, written over 3,000 years ago, and drew primal insights from the ancient meditative sources to formulate their quantum theories.

In modern scientific history, it wasn't until quite recently that the existence of atoms was proven experimentally, when the British scientist J. J. Thompson in 1897 accidentally discovered electrons, one of the three primary building blocks of all atoms. In 1911, Ernest Rutherford and his colleagues proved that atoms also have nuclei. This discovery happened around the time Einstein was using mathematics to explore his new vision of relativity.

This notion of the existence of atoms (and anything smaller than atoms) has been around in Western culture for just over a hundred years. Only during the last fifty years or so has the subatomic quantum vision of reality exploded into common view. Don't feel behind the times if you're still struggling to comprehend what the bleep this is all means. The quantum mindset is still in its infancy; we're all pioneers exploring the new terrain of subatomic science, just like Magellan explored the mysteries of a (hopefully) round planet.

Where did all these atoms come from, and how were they manifested? In the partly-proven scientific myth of the Big Bang "moment of creation," which seems to have happened around fourteen billion years ago, we're told in the first micro-seconds, free-flying electrons and protons came into being, but no one knows how or why. The primary mystery of creation is still a gigantic unknown.

The *National Geographic* article, "*Every Black Hole Contains Another Universe?*" states a new "multiverse" mathematical model, "According to the new equations, the matter that black holes absorb and seemingly destroy is actually expelled and becomes the building blocks for galaxies, stars, and planets in another reality."

One way or another, our universe began with the tiniest building blocks which would come together to generate all the matter in the universe. Scientists estimate that just one micro-second after the Big Bang, the universe was the size of a golf ball and consisted of six subatomic particles: neutrons, protons, electrons, anti-electrons, photons, and neutrinos.

As we'll see later, these tiny particles were made up of even tinier energetic bundles of quarks, gluons, and electrons. Incidentally, there seem to have been no radiant light particles in the universe at this time. The phenomenon of light would begin 500 million years *after* the Big Bang, after things cooled down and stars came into being and began emitting light photons.

Around 380,000 years after the Big Bang, electrons began to be trapped in orbits around protons and neutrons, forming the very first atoms, mainly helium and hydrogen, which are still by far the most abundant elements in the universe. Through some mysterious force of nature, these electrons and protons, with their opposite charges (negative and positive) were attracted to each other.

Here we observe, with the first atoms, the original instance of the mutual attraction of opposites. This core bonding dynamic of balanced and opposite-charged particles reveals one of the primary features of our universe: the phenomenon we call symmetry or energetic balance, where equal charges balance each other and maintain enduring relationships. The relationships in atoms is so powerful that most of the elemental matter formed 14 billion years ago are still bonded—wow!

Right now, you are entirely made up of these atoms that were formed 14 billion years ago. The same is true of the people you love and everyone else on this planet. It's not a cliché to say that we are all stardust. When we include this fact in our expanding world view, we can see each other in a bright new light. We're all 14 billion years old! In terms of your emerging quantum mindset, this is a core revelation.

That isn't the end of the story from the quantum-mechanics perspective. What in fact is the nature of the primal binding/bonding force that's lasted 14 billion years, still maintaining primal energetic relationships, that sustain your current physical and energetic presence?

THE GLUE THAT BINDS US ALL

We now come face-to-face with one of the continuing mysteries of even the most advanced quantum physics: What is the force that holds negatively-charged electrons to positively-charged protons? No one knows the answer, not even now. It's easy to say that it's "the electric force," as your high school or college teacher probably told you, but no one knows the origin or cause of that electric force, just as we don't really know what electricity is.

Science has found it's possible to manipulate matter and energy at the atomic level and accomplish remarkable goals (think atom bombs), even without comprehending the deeper mystery of the phenomena. In this same spirit, electrical engineers develop more and more sophisticated nanocomputer systems without knowing what electricity is.

One advantage to building things without understanding them is we can do just that—build new things that help us in life. The downside of plunging blindly ahead, as we are, is that we just might blow ourselves up or run the world's civilizations right off the proverbial cliff.

As an introduction to the power or force that holds the nucleus of an atom together, we need only think of the atom bomb, where a single free neutron strikes the nucleus of a radioactive atom (uranium or plutonium) and bangs a few neutrons free. As these neutrons split off from the nucleus, they release a vast amount of energy—that primal binding force! Then they strike the nuclei of other atoms, splitting them and releasing more energy and more neutrons in an almost instantaneous chain reaction.

This sobering reality leads us naturally to ask what the proton is. How is it possible for such a tiny thing to pack such an energy punch? The answer to this core question began in 1918 when the Englishman Ernest Rutherford demonstrated that at the center of the hydrogen nucleus there was both a proton with a positive charge, and surprise! There was also a chargeless particle in the atom's nucleus. Furthermore, this neutron was strangely bonded to the proton by some undiscovered force, even though the neutron had no electromagnetic charge at all—zero!

Scientists now knew the three primary particles of the atom: the electron, the proton, and the neutron. But what was holding protons and neutrons together in the very center of the atom? Your teachers probably told you there's a simple answer to that question—there exists yet another quite different and equally mysterious force called "the strong force." This force is considered the most powerful force in the universe, but it only functions at very tiny distances, like in the center of an atom.

Here's another very strange phenomenon: The proton and the neutron are held together by this massive invisible force, and they are intimately engaged with each other, but they don't ever touch or merge. Instead they maintain a discrete tiny distance, acting as separate entities united in the essential cause of sustaining the tiny mighty atom wherever it's found in the universe.

You might want to pause to imagine the proton and the neutron in an atom, feeling how they're so strongly attracted to and bonded with each other—but also remaining distinct and different. Now think of a relationship you're in or would like to be in, where you're deeply involved and yet quite separate ...

If you're over forty, your physics teacher probably didn't go into too much detail about this proton-neutron relationship, because until just recently not much was known. Before the quark and other even-tinier elements of the atom were discovered, scientists assumed we had hit the smallest particles possible, the ultimate building blocks.

The general theory was that gradually those three atomic building blocks of electrons, protons, and neutrons became more complex atoms as stars began forming to generate very high temperatures and pressures. This enabled more complex atoms to be created with more and more protons and neutrons in the center, with parallel electrons in orbit.

As the universe continued to expand, dozens of new and bigger elements were created until there were almost a hundred distinct elements, each with a different proton count at their center and an equal number of electrons and neutrons. Beyond that naturally-occurring collection, no more elements existed until humans came along and discovered how to artificially and temporarily create a few more—24 to be exact.

By the 1950s, scientists had concluded that all matter throughout the universe consisted of the 90-some primary elements. They also concluded there are four original types of invisible forces that hold absolutely everything together in harmonic motion: a strong force and weak force found only in the atomic nucleus, the gravitational force, and electromagnetism. No one knew what these forces were or from where they drew their power. They were somehow just eternally and everywhere present in the universe.

What became experimentally certain was that most particles are charged with an energy or force that either attracts or repels other particles. Furthermore, each charged particle or collection of bonded particles maintains a particular charged electric force field around it. This force field, again quite mysterious and invisible, radiates outward in all directions and affects other particles, even at a distance.

We can't yet explain the source of the primary forces that energetically hold everything together. There are vast mysteries in science that simply remain mysteries. We'll see a bit later that the wave radiation of a particle or force field has now been demonstrated experimentally to radiate outward, even to the very edge of the universe. Specific to your personal relating, your body likewise has a charge, your personal resonance, that also radiates outward.

You might want to pause for a few breaths again after reading this paragraph. Close your eyes and allow the reality we're exploring to permeate your focus, so that your mindset can expand to include this notion that right now, your organic presence is energetically being broadcast outward in all directions ... just see if you can sense how your ongoing presence is touching everyone around you with an energetic radiation that science has now proven exists, even if we're not aware of it with our usual senses ...

THE MICRO PLOT THICKENS

We're dipping fairly deeply here into detail, but hold with me—because as you build a clearer model in your mind about the fundamental way that things attract in the micro universe, you're strengthening your inner sense of how you likewise relate with the world around you.

These four forces of quantum physics are the exact same forces that move you through life, and they're utterly important!

Until the early 1960s, scientists were certain they had identified the ultimate tiny units of reality with the protons, neutrons, and electrons that made up every atom in the universe. Then at Stanford University, a physicist named George Zweig and his colleagues discovered each proton and neutron in every atom could be subdivided into even tinier particles which they named quarks.

With this discovery was born the academic division called subatomic physics. But what were quarks? To this day, no one has seen a quark; we only assume they must be there, based on substantial amounts of research showing their influence on their environment. We now know that all protons are made up of two "up quarks" and two "down quarks," whereas neutrons are made up of two "down-quarks" and one "up-quark."

Quarks, unlike other particles such as electrons, photons, neutrinos, and so forth, are found only in the nucleus of the atom. Quarks have been found to behave both as distinct particles of matter and also as energetic waves. They seem to exist as wave functions except when they are being observed.

This is one of the most important realizations in scientific theory: the emergence of the idea that all matter in the universe is made up of energetic waves which appear as solid particles only when disturbed through external observation. This is very important to remember. Everything is an energy wave until its observed. Furthermore, Werner Heisenberg insisted that it is impossible to observe or measure a wave function directly, because of what he called the uncertainty principle, at any given point in time, a wave packet's location is unpredictable. This unpredictability is linked to our freedom of choice because every wave is a range of possibilities until it's observed by consciousness.

Furthermore, Heisenberg and fellow physicists concluded that science will never be able to see or comprehend the full nature of matter and energy, because of the unpredictability of the energetic waves that permeate the universe and constitute reality. The use of imaginary numbers in quantum math also expresses the perennial mystery that lies beyond the illusion of solid matter.

Regardless of the illusiveness of the wave patterns that seem to make up quarks, scientists have continued to delve deeper into the nature of quarks. In so doing, they've discovered yet another deeper level of energetic relating that seems to glue everything together. Let's take a quick look at the dynamics of this ultimate glue-energy.

BEHOLD THE GLUON

We're now getting down to the true nutty-gritty of this whole exploration. Research shows that quarks are always positively-charged with the strong force, similar in this aspect to gravitation. Recently, scientists discovered there are actually three distinct types of positive charge in the strong force of quarks, and these three different charges are determined by some super-mysterious force or energetic bundle called the gluon, which was discovered in 1979 at Germany's PETRA collider.

Gluons and light photons are very similar. They travel at the speed of light (gluons sometimes travel more slowly), neither has mass in the classic sense, and neither has a positive or negative charge. Similar to photons which are the force-carriers or converters of the electromagnetic field, gluons are the force-carriers of the strong-force field.

Gluons exist naturally only in the nucleus of atoms, and somehow they channel or possess or generate or convert the illusive strong force that holds an atom's nucleus together. Gluons somehow glue quarks together to form protons, and they also give mass to the quarks that make up the proton.

There are eight different types of gluons, and each has its own anti-gluon. In the spirit of subatomic symmetry, there are also anti-protons, anti-quarks, and so on. And gluons can link with quarks and also with other gluons. The story gets ever more complex, but what's important for our discussion is that there's a remarkably strong attraction happening, via the converter gluons, within the nucleus of all atoms. When you try to pull bonded gluons apart, the farther apart you pull them, the greater the force that's holding them together. It's the *inverse* of the inverse law, where a force becomes less as the distance increases.

Here's another remarkable discovery: The binding force of the gluon is responsible for some 99% of the mass of both protons and neutrons, but as I mentioned earlier, gluons themselves don't have any mass at all. As Professor Mindell states, "Particles are a magical explanation that depends on a world view that believes everything is reducible to a thing. But science has now led us to where we can deduct that at the core of any particle we find coded energy, not anything solid."

Even if we speak of an atom as having mass, when we dig into a molecule of "solid matter" and encounter a single atom of that molecule, we discover that the mass in that atom fills only about a thousandth of one percent of the space the atom occupies—again, mass is a very tiny tiny bit of what we consider the volume of an atom.

In your emerging quantum mindset, this means that your own body is mostly *nothing*. When you push one finger against another, what's happening is that the electron cloud around each atom is insisting on maintaining its "orbit" around its nucleus and pushes strongly against any intrusion.

When you feel a light breeze, individual air molecules are pushing gently against your face and flowing around your more-solid skin. Take something like a stone with more mass than air (many more protons in the nucleus of each atom), and the heavy load of protons and electrons can put a dent in your skull, even though the rock itself is mostly just air!

Gluons possess energy, but they do not possess any mass. Think of it this way: if $E=MC^2$, and a gluon is Energy, then that Energy is also equivalent to Mass, as Einstein said. Gluons don't seem to linger much in their potential Mass state, and the amount of energy in a single gluon cannot be isolated or calculated. All science can say at this point is that this "force carrier" between quarks enables quarks to exchange the strong force, and this somehow generates a bunch of quarks that in turn make up what we call the proton(s) in an atom.

Somehow the strong force manifests its presence and power through the gluon, which then manifests the quark, which in turn manifests the proton and neutron, which manifests (along with the electron) the atom, which bonds and manifests the molecule which manifests all known matter in the universe. Yes—we got it!

Physicists have found twelve elementary subatomic particles that serve as the building blocks of all matter: six quarks, as mentioned above, plus 3 electrons (electron, muon, tau) and three neutrinos (e, muon, tau). Quarks are around a hundred times heavier than electrons, and gluons, quarks and electrons are "point particles," meaning that they don't take up any space at all.

Really, how tiny can things get? Perhaps we've now finally reached the historic point in physics where no smaller particles or energy bundles will ever be found—but I'm not holding my breath.

Speaking of breath, how about another pause to let all this information percolate through your mind ... see what happens when you first tune into your breathing and let the dust settle ... be aware of your whole body here in this present moment ... considering everything we've just talked about, tune into whatever feelings arise deep inside you, right in the middle of your breathing ...

If the scientists have it right and your own body is made up of these tiny quantum building blocks of intelligent energy ... what does it feel like to have a pure-energy presence? Do you feel your quantum self now becoming empowered to set an intent ... and manifest what you feel you want and need?

Chapter 5:

STRINGING IT ALL TOGETHER

There's one final layer of quantum theory to share with you before we finish off this first section of the book. This theory attempts to bring everything together into a grand Theory of Everything (TOE), similar to Einstein's search for a unified field theory, a single model that aims to explain the fundamental interactions or mechanics of the universe as a whole. Many prominent physicists adhere to what is now called "superstring theory" or the M-theory, which purports to mathematically describe the geometry of space.

String theory states that at the very beginning of the universe, the four fundamental forces we've been talking about were a single fundamental force. This force expressed itself in the universe as ultimately-tiny vibrating strings or looped strands. Various combinations of vibrations or specific oscillatory patterns expressed in the strings generated the different primary particles with their unique mass and force charge. An electron in this model will be a type of string that vibrates in a certain way, while each of the eight types of quarks will be a type of string vibrating in a slightly different way.

Einstein had already expanded classic 3-dimensional reality to include the fourth dimension, time, and thus created the concept of spacetime, his unified 4-dimensional matrix of the known universe.

This "general relativity" theory presented a beautifully symmetrical vision of how everything in the universe moves.

Even in 1916, when Einstein presented his general theory, the entire known universe consisted of just our Milky Way galaxy, and this known universe was considered to be static, neither expanding nor contracting. That's the mindset and world view that my grandfather carried around in his head; there was nothing beyond the Milky Way.

Einstein's math startlingly predicted that the universe was in truth not static and unmoving at all, and this prediction was proven in 1990 with the employment of the Hubble telescope and the discovery that the universe is vastly more than the Milky Way. It also proved that the whole universe is expanding. This revelation in itself called for a great quantum leap in the average world view.

Around this same time, mathematicians predicted (and recently astrophysicists have located) a supermassive black hole at the center of our galaxy. Soon it was found that most galaxies have their own black hole, or even more than one, and this explained how galaxies and new stars were continually being formed and destroyed.

Imagining a black hole of gargantuan size has continued to stretch our imaginations as our world view has likewise expanded. Very recently, astrophysicists concluded there are literally millions of smaller black holes shooting across our galaxy and gobbling up interstellar gas. Furthermore, new evidence indicates that black holes could be portals into other universes—but more on all that later.

In the 1930s and 1940s, using Einstein's math, scientists predicted that according to general relativity theory, galaxies should all be falling apart, and the universe should be collapsing. Physicists went back to the proverbial drawing board and came up with the startling idea that there must be something called "dark matter" filling the universe, in order to make everything function as observed. In the 1990s, physicists realized that for all the dark and regular matter to be accelerating, some vast field of undetected energy must also exist, to account for the acceleration push.

Dark energy is unique in that it's considered to be uniformly distributed across space, which might help to explain what's called particle entanglement where two particles somehow communicate instantly with each other, even across vast galactic spaces. The presence of dark matter and dark energy existing where we detect only empty space could also account for such things as telepathy, synchronicity, distance viewing, and many other supposedly-supernatural phenomena (more on this later).

Here's another staggering statistic: For everything to be as it is observationally in the universe, around 95% of the universe *must* be made up of dark matter and dark energy, which means that the known universe's composition of regular matter represents only about 5% of the universe's total composition. As with your own body and the atoms that it represents, the universe is mostly a mystery of non-material space and energy. How does it feel to have a body that's alive and well in a universe that's mostly made up of some energetic substance we know nothing about?

We've already mentioned how photons, gluons, and electrons, the energetic connectors of the universe, function as both particles and waves. This was proven in 1927 with the famous double-slit experiment. In Heisenberg's probability theory, such waves or quantum particles possess what's called superposition, which means that a particle at any given instant can be anywhere within its operational field.

Dealing with this inherent loss of predictability at base levels of reality jolts most people, forcing them to shift entirely beyond Newton's confident predictability into a quantum state of utter uncertainty.

This quantum jolt has reverberated throughout the modern world and is at least partly responsible for our current era being called the age of anxiety.

To integrate all of that, let's focus again on the specifics of string theory, where all the quantum point-like particles we've called the smallest units of matter have now been replaced by even-more-tiny one-dimensional energetic bundles called, for lack of a better word, strings. These miniscule loops of flowing energy are said to propagate throughout the entire universe, where they interact energetically and intelligently with each other. In this theory, these strings are all there is. All objects and interactions in our universe are made up of these vibrating energetic filaments and membranes.

Now for the whopper: string theory insists that there must be at least ten or eleven spacetime dimensions, rather than just the 4-dimensional spacetime reality of Einstein. The math for this theory only works when there are that many dimensions. Also, string theory allows for more than one universe, a theory which seems to be gaining momentum recently.

The most valuable takeaway from string theory for most of us is that everything in the universe, in fact the universe itself, isn't made up of things, it's made up of pure energy. The way these tiny packets of energy manifest everything we think of as real, as matter, is through the almost infinitely-complex vibrational interactions and relationships between these super-tiny energetic strings.

Here's a concise summary of all this: "The vibrations of these strings generate everything out of a vacuum, such as the different characteristics and features of subatomic elementary particles and all elements of the periodic system. Ten and more dimensions are needed to describe this reality. All kinds of particles are unified by this Super String Theory because each particle differs only by the oscillation pattern of a string. Even space and time are supposed to be subject to the vibrations of these strings."—"*The Power of String Energy*" discussed by Dr. Henryk Frystaki, *Science Illustrated.*

This most recent plausible theory of how the universe works is grounded on resonance, on vibration, and this is important in the evolution of quantum science. As we'll explore in more depth later, one of the key attributes of humankind is that we've got rhythm. We have a sense of vibrational expression. Every tribe and culture in recorded history has had music as a primary unifying experience. This new leap of quantum physics into vibrational modalities to explain reality somehow deeply excites me. Stay tuned.

> *String theory, with all its resonance dimensions, is just that—a theory. Theories are models that hopefully reflect our reality. The question is: Does the string theory, with its vibrating energetic packs of ultimate building blocks ring true somehow for you? Harmony is the essence of successful relating and community. Vibrations are the medium through which we receive sensory information. So, are resonating strings what the universe runs on? We talk about our heart-strings … you might want to pause, put the book aside, and breathe into whatever feelings and insights rise upward from heart to mind right now …*

EXPANDING YOUR WORLD VIEW

If energetic coded waves of some sort are the quantum medium we all live in, what does this tell us about how best to approach changes? People often talk about good vibrations, about someone having "good vibes." Sometimes we have high-frequency days and sometimes low-vibe days. When someone talks to us, their sound vibrations penetrate deep into our minds and bodies. Light rays warm the skin and fill our minds with information and brightness. Everything seems grounded in vibration. Our microwave ovens, our wi-fi and internet, our phones, and all the rest run on waves and vibration.

We know that if we focus on putting out good vibes, we tend to also feel good in our own skin. If we want to attract someone to us, we need to send out good vibes in their direction which they might pick up and respond to. Most of us can admit we live in a mysterious vibrational world, even if we can't fully comprehend gluons and string theory's resonant foundation.

We also know that personally we're often caught up in low-vibration moods and disharmonious thoughts and feelings, even though creativity and successful communication require a high-vibe inner charge. Enthusiasm and optimism make things happen but often our awareness is dominated by thoughts and attitudes that make our emotional resonance dark and low and disharmonious. In a universe where everything is energy, we seem to manage our allotment of energy in ways that often don't really serve us.

So ... what stands between us and manifesting the optimum life seems to be, well—our own current mindset and world view. How we see the world and our ourselves, positively or negatively, definitely determines our energy level and vibratory condition. We all know people who have a depressive, pessimistic attitude toward life, and we know they suffer from their negative mindset. We also know people who are habitually bright and positive and are busy enjoying life to the hilt.

It's a fact of life that we are all programmed as kids to feel like our parents felt. And negative events in life can definitely leave us with PTSD-style depression and anxiety.

The most astounding facts about adopting beliefs and traits from our parents is that we not only adopt their beliefs, but also their fears, and traumas, some even before we're even born. Among the tens of thousands of people who were directly exposed to the 9/11 terrorist attack were approximately 1,700 pregnant women. Many of these women developed symptoms of PTSD, and some of their children inherited the trauma that their mothers experienced on that horrific day. Some transferred trauma to their unborn children, and their babies were born with noticeably higher cortisol levels than normal.

Generational trauma is a real thing and science agrees. I'd like to refer to a study I read in the *Washington Post* by journalist Meeri Kim:

> In the experiment, researchers taught male mice to fear the smell of cherry blossoms by associating the scent with mild foot shocks. Two weeks later, they bred with females. The resulting mice pups were raised to adulthood having never been exposed to the smell. Yet when the critters caught a whiff of it for the first time, they suddenly became anxious and fearful. They were even born with more cherry-blossom-detecting neurons in their noses and more brain space devoted to cherry-blossom-smelling. The memory transmission extended out another generation when these male mice bred, and similar results were found. Neuroscientists at Emory University found that genetic markers, thought to be wiped clean before birth, were used to transmit a single traumatic experience across generations, leaving behind traces in the behavior and anatomy of future pups.

These 9/11 babies and newborn mice were born innocent to the world around them, yet they harbor generations worth of information, both bad and good, passed down by their ancestors, the bad and the good.

Since these studies prove that when significant trauma is inflicted, it's passed down through generations; consequently, this would also mean that when we heal ourselves, we can also heal our entire generations. Thus further adding to the notation that we are all intimately connected to one another beyond space, time, and even throughout lifetimes.

But are we hopeless victims of the world view and energies we inherited, or can we now take charge of our minds and emotions and evolve in more resonant, harmonious, and positive directions?

Without question, we all have the capacity to transform our mindset. But to be specific, which life variables can we work with to change, heal, and advance our mindset to accomplish our desired goals? We're seeing how science is all about variables, and change of any kind is based on which variables are highlighted and which are pushed aside and ignored. And if you want to take that leap into a brighter quantum, you'll need to identify the mindset variables that you can improve or transform to your advantage, and then act to do just that.

THE THREE VARIABLES

Let's end this rather heady science chapter by considering the variables you have inside your own mind and world view that you can affect in positive directions. Let's hold throughout to the underlying scientific premise that you'll do the best when you align your personal actions with the quantum laws of the whole universe. This conscious alignment is the key to both empowerment and fulfillment.

The adage, "No one fights reality and wins," expresses this in traditional terms, and points us in the direction of developing a mindset that's resonant with the most up-to-date reality-based science. The ancient biblical saying, "Not my will but thine O Lord," certainly implies that we should align our personal intent with the core laws (will) of the universe if we want to pack true manifestation power.

In this act of aligning your personal intentions with the higher order, process, and power of the universe, there are three basic inner variables that you can use to expand your world view and improve your everyday experience. Let's look at each of these in turn, and then end the chapter with a basic process that will actively help you advance in your alignment and improvement quest.

Realistically, what can you change, and how can you accomplish this change? This is pretty straightforward: 1) you can change what you do out in the world (the physical dimension); 2) you can change how you think (the cognitive dimension); and 3) you can change what you feel in your heart (the emotional dimension). Anything else?

Perhaps you might also have the idea, "I can change my intuitive spiritual engagement with the higher integrative wisdom, intelligence, and force in the universe." This is a great lofty high-order intent, but I think you'll find when you deal with the first three, you also deal naturally with this fourth dimension. I've found that when a person develops an expanded quantum mindset and world view, the fourth level of higher-order alignment and empowerment happens as a logical consequence. I'll leave that up to you to decide.

So your challenge is to engage with the physical, mental, and emotional dimensions of your personal life, with the intent of raising your core vibration, expanding your quantum world view, and awakening your deeper heart-empowered capacity to align with the higher laws of the universe ... and thus create the life flow you yearn for.

But—what's the proper order to approach this? Again, logic is fairly clear about this order. First, tune into what you truly want to accomplish, what your driving intent is, and what your deeper needs are. There's thinking involved in this first process, but your emotions and deeper intuitive function will let you know when you're realistically aligned with your core intent.

Then you need to go to work on your mindset, apply your mental powers, put aside old mental habits, attitudes, worries, and anything that hinders your progression. Expand your world view so your enhanced vision of your potential matches your intent. This is practical work done in your mind—clearing your subconscious of negative one-liners you were programmed with as a child, so you're free to take the leap into newness.

The third step in this basic manifestation-fulfillment process is to begin interacting with the world around you so that you manifest the new people, things, and situations you feel drawn to, attracted to, and somehow already positively entangled with. If you complete the first two steps successfully, then the third action step should mostly come of itself, as you allow the primal quantum forces we're exploring in this book to do their thing energetically and bring into your life what you're naturally attracting.

Let's jump in with the first variable: setting your intent. Rather than deciding on something specific that you want to get in your everyday life ("I want to find true love ... I need a new coat ... I want a better job ..."), let's first begin to clarify your underlying impulse toward positive change. We'll be returning to this focus throughout this book and program, because you'll find that your core intent like everything else, is constantly evolving, expanding, and becoming more empowered and clear as you change.

What Is Your Primary Intent?

As you read these words, let your awareness expand to include your next inhales ... and exhales ... just settle into a relaxed feeling throughout your body ... become aware of your living presence right now in this moment as you breathe ... feel how you're balancing in the earth's gravitational force field ... and feel how every cell in your body, every quark in every atom, is vibrating with life, with power, with intent, and with contentment ...

Let yourself begin to tune into whatever deeper feelings you have in your heart ... what do you feel good about in your life ... where do you feel pleasure ... challenge ... satisfaction ... meaning?

Allow yourself, without any judgment, to notice where you feel a yearning ... a desire ... a hunger for something new and currently missing in your life ... what do you want to attract into your life? ... just feel it out there, waiting for you to develop an attractive charge that will bring it into your personal bubble ...

Now see how you might complete the following statement: "In harmony with my deeper needs and intentions, I want to begin to attract (such and such) into my life ..."

ONE-LINER PROCESS

Along with that first foray into clarifying your overall yearning and intent, let's end this chapter with a beginning exploration of how to identify and put aside the negative one-liner rules and beliefs you might be carrying in your mostly-unseen subconscious mind. In this book, we're using the terms "subconscious" and "unconscious" interchangeably, rather than differentiating between the two words like Freud and Jung did.

Every person inherits an initial mindset and world view: a set of short fear-based assumptions from before birth onward throughout childhood. Some of these rules and assumptions are correct and of great value and need to be honored and sustained, but many others need to just be dumped as dead weight.

You can't fill a cup that's already full. There's a whole academic field called Positive Psychology that deals more and more effectively with the notion of "de-beliefing" old defensive rules and self-defeating assumptions picked up in childhood. If you pay attention to the sub-liminal thoughts that flow through the "back of your mind" you'll probably find you're often talking to yourself in ways that knock you down rather than raise your vibes. Why is this happening, when it's so obviously counter-productive?

Your current adult world view is a natural product of your genetic inheritance, cultural and religious upbringing, the experiences that impacted how you see yourself and the world, plus all the reading and discussing and reflecting that generated a sense of meaning in your life. These ingredients have coagulated into your current world view.

You've probably found that even when you try to be positive and strive to get ahead in life, somehow there's always a depressive subterfuge that keeps telling you you're no good, you can't make it, being rich is bad, or you're too dumb or weak. You have literally been programmed with such one-liner beliefs, usually through mental and emotional osmosis, absorbing the negative attitudes you were immersed in and inherited from your family and community.

Throughout the rest of this book, we will return to this theme of "de-beliefing" so you can consciously identify your own negative one-liners,—and discard them in favor of positive one-liners that will support you as you leap into a quantum mindset, rather than continue to subtly defeat you.

Let's run through several debilitating and too-common one-liners to clarify what they are. As you read each negative one-liner, think of its opposite. What statement of positive belief can you program your subconscious mind with to override the negative?

The more you say positive things to yourself, the stronger the statements will become in your life. Each statement carries power, and you tap into this power by repetition, by consciously talking to yourself and telling your unconscious what you now choose to believe, based on the level of reality you're now seeing to be true. This process works both for integrating quantum-science facts and truths you're encountering in this book, and for all other aspects. Talk to yourself and tell your unconscious mind what you choose to believe now!

Read each of these one-liners to see if it's a belief or attitude that you hold in your subconscious mind. If it is, try to come up with an equal and opposite one-liner that you can establish as your new rule, belief, or attitude. Remember this is just a sample list. You might notice your own unique one-liners as they run through your mind and create a new positive one-liner to install as a quantum-leap replacement.

"I don't deserve success."
"If I'm honest I'll get in trouble."

"People are so mean!"

"No one cares about me."

"Having friends always gets me hurt."

"I'm worthless."

"It's all going to hell."

"I'm not good enough."

"Having money is bad."

"People can't be trusted."

"I can't handle it."

"I always fail."

"Life is terrible."

"I don't know what I want."

"I don't have time."

"I can't keep up."

"You can't trust strangers."

"I'll never understand."

"I don't need anybody's help."

"If you want it done right, do it yourself."

"Life sucks!"

"I'll never get ahead."

"It's not fair!"

"I don't deserve anything."

"Life is so depressing."

"Nobody likes me."

"I'm just dumb."

"It's all my fault."

"Feeling sexy is a sin."

"I'll always be poor."

"The cards are stacked against me."

"It's all just hopeless."

"Money is the root of all evil."

"It's too hard."

"Something bad is going to happen."

"I can't do anything about it."

"I don't believe I can."

"Trusting is too dangerous."
"I'm too old now."
"Life is a constant struggle."
"I'll never get ahead."
"Being rich would ruin my life."
"I'm afraid to change."
"The world is going to the dogs."
"I have to just keep fighting."
"I always get hurt when I love someone romantically."
"I just don't have what it takes."
"It's too late to do anything."
"Nothing ever changes."
"Life's too much for me."
"I don't know where to start."
"I'm all on my own."
"Being nice is for sissies."
"People always take advantage of me."
"I can't make it out there in the world."
"It's too risky to try."
"My feelings always get me in trouble."
"People always disappoint me."
"I can't do this."
"The whole world is against me."
"I give up."
"I can't trust my heart."
"Everything's going too fast."
"Trying new things gets you in trouble."

Whew—that's a lot of negative one-liner attitudes to take in at once, but most people have loads of them. Notice they're only one sentence long. Why? Because you mostly adopted these attitudes and assumptions when you were just learning how to talk, and you first learned to talk in one-liners. To effectively reverse these downer assumptions, you'll want to use one-liners too. Fight fire with fire.

What we're doing here is acting to clear your mind of old inherited fear-based reactions to life. Parents naturally want to protect their kids from possible danger by scaring them away from situations that might hurt them. Most parents simply do their best and blaming them is counterproductive. After all, they were programmed with their parents' fears and prejudices. The link goes back many, many generations. What we're doing now is breaking this fear-based belief chain, so that you're free to leap forward.

You can do this same process with just a single word as well. The same word can have different meanings to different people. Take the words good, bad, humility, trust, hate, sex, church, evil, commitment, purpose, and God. These are what I call loaded words that can be connected to something deeper within your subconscious. What do these words mean to you? Why does a particular word make you feel a certain way?

Write each of them down and give your own meaning to them.

Negative one-liners and lack of forgiveness are energies that hold you back in life. Unless you learn to reverse your negative one-liners and practice forgiveness, you will feel like you are swimming against life's currents.

TRUSTING IN THE NEW

Fear makes a person contract, while trust helps them to expand. Staying stuck, feeling depressed, hopeless, anxious, angry, or defensive will directly inhibit your expansion into a wider, more creative mindset. The new quantum world view you're exploring in this book cannot be embraced if old outdated assumptions dominate your mind. It's essential to replace all the "outdated beliefs" and "fear-based attitudes" that might get in the way of dreaming bigger dreams and manifesting at higher levels.

The good news is that when you get rid of a negative one-liner and replace it with a more realistic and positive one, you often get rid of all the similar negative one-liners you have. Most of them are related. All of them seem to be associated with feelings of inadequacy, self-doubt, vulnerability, fear, and weakness.

What happens to your energy when you consciously confront a negative one-liner and realize it doesn't serve you and doesn't reflect your present reality? You pull the plug on the energy that negative attitude needs to survive. When you replace it with a positive belief or assumption that rings true, at energetic levels you're taking your power of attention away from that realm of fear-based thinking.

Your power of attention is just that—it's an energizing force. What you focus on, as we're seeing in more and more depth here, is what you materialize and what you manifest. As long as your power of attention is being sabotaged by negative one-liners, you will lack the power to focus on and manifest anything positive.

The great British poet T.S. Eliot once said, "To make an end is to make a beginning. The end is where we start from." Psychologically this is absolutely true; we must find what is holding us back from what we desire and act to put an end to it. Only then can we make a new beginning. Cleaning house is required. Clearing some space in your mind by throwing out debilitating beliefs makes it possible to create a mindset that empowers you.

There isn't one standard quantum mindset that fits everyone. You developed a unique world view as you were growing up, and now you are developing a fresh world view that moves beyond stereotyped attitudes and assumptions into you own true uniqueness. You might have been programmed with general racial stereotypes and cultural attitudes, and educated with the same textbooks and teachers, even brought up by the same parents in the same house, but as all siblings know, you're an utterly unique conscious entity.

As you develop an expanded and transformed mindset based on new information, you'll gain even more freedom to develop your uniqueness. This power to recreate your personal mindset in whatever way you prefer is golden and too seldom tapped.

Consider the newly-proven fact that the universe is not just expanding, it's also constantly accelerating. The whole world you live in is caught along with you in this sense of non-stop expansion, of everything speeding up rapidly, going faster and faster into the new. Most of us these days are suffering from future-shock of one kind or another, struggling against the "free floating" anxiety we feel because our one-liner is "I can't keep up." Over 40 million Americans currently have "anxiety disorder"—it's an epidemic. The future is so scary that we just freak out and contract rather than expanding and dealing with the challenge.

In the Newtonian mindset, this "acceleration of everything" causes anxiety because we have no idea what's coming next, and we fear losing control of what's coming at us. We lack trust in the future. But within the quantum mindset, we have the larger assurance that the entire universe is unfolding right now within the perfectly-balanced laws of the Creator. We can trust that acceleration is okay.

As we're seeing, the very foundation of life in this universe exists within the dynamic of accelerating change. The universe has been accelerating for 12 billion years. Life didn't cease to exist when we found out we weren't the static center of God's creation. Life will continue now that we're discovering that acceleration is a primal quality of the whole universe, us included.

Therefore, we must know, deep in our bones and atoms and quarks, that our nature involves acceleration and constant change. The universe seems to be under the control of a deeper pervading "wisdom of life" and that we can relax and trust in it. Balance, symmetry, and perhaps even the power of love seem to rule both our genes and our stars. I don't say this in any religious sense; I say it quite sensibly as a statement of the new trustworthy hope that's emerging from the quantum world view.

Consider how you relate with people. The old world view positioned you as competing for scarce goods against hostile agents—you against them, long live tribalism and all the rest. The quantum world view shows that you're not actually an isolated individual fighting against the whole world for your solitary survival. In the quantum world you're by definition intimately and vibrationally related with every person around you.

Ecologists have proved statistically for decades that there are plenty of resources to go around for all. "We live in a world of scarcity," is a negative one-liner that undermines our ability to shift into an economy that is wise enough to distribute goods fairly. As long as we believe in scarcity as our economic model, we'll continue to generate scarcity. If we bring the belief of abundance into our mindset and world view, we'll transform the world situation for the better. This is the power of the mind to manifest a better world.

It's the same with relationships. If we believe people can't be trusted, greed dominates love, or we're isolated entities in a hostile world, such negative one-liners will manifest in our relationships or lack of them. In quantum physics, the truth is that we're at so many levels always entangled with each person we interact with, and we have the power to tune into subtle shared resonances and harmonics. Unfortunately, if we don't believe the new science, we won't experience being in sync at core levels with the people around us.

Holding us back from leaping into this new sense of harmonic engagement with the world are our old outdated world views, our ingrained prejudices and unchallenged fears, and negative one-liners that cloud our realization of who we really are.

Rather than staying stuck in old limited reflexes, thinking the same negative thoughts that lead to the same limited outcomes, we're discovering the freedom to shift our perspective and see new alternatives, then choose to do something new. The new world might seem less predictable these quantum days, but with Heisenberg's uncertainty principle we also rest assured that each new moment offers us many new choices and opportunities.

In this spirit, let's end this chapter by going deeper into the process of shifting one-liners to your advantage. This guided session will show you a repeatable process to choose new rules, positive one-liners, and realistic beliefs that will in fact set you free.

CHANGING YOUR OWN RULES

The blunt truth we're learning is that unless you act to consciously identify negative beliefs and discard them in favor of more functional and positive one-liners, you remain a victim of these subconscious attitudes and assumptions your entire life. Your ingrained world view and subconscious one-liners could be the only aspect in your life that doesn't change, unless you become passionate about breaking free from your negative programming and actively apply a "de-beliefing" program (like the one you'll learn in this book) to transform your mindset in the direction you prefer.

Let's experiment with a simple but powerful method to help you leap into an expanded quantum mindset. We'll continue advancing this method throughout this book and program, so you have a clear path to follow as you develop your new world view.

Exercise: Transforming Mindset Assumptions

First, let's pause and tune into how you're feeling after reading this chapter. New ideas stimulate all sorts of feelings; that's one of their main roles. These feelings can either provoke you into action, lull you into contentment and joy, make you feel anxious or worried, or excite you with realization and delight.

Your feelings are what move you, and movement is a core function of the universe. When you feel positively moved and, when feelings flow freely through your body, you feel energized, tuned in, and turned on. In this high-frequency state, you sense newness and opportunity, even in the very air you're breathing. In this inspired mood, you can manifest! Let's begin with how you feel right now:

> *To instantly tap into your inner feelings, take charge of where you're focusing your mind's attention right now, and aim that attention toward the physical sensation of the air flowing in and out of your nose ... feel your chest expanding as you inhale ... and your belly muscles tightening as you exhale ... make no effort to breathe, just allow your breaths to come and go all on their own ...*

Now let's delve deeper into how changing a one-liner can transform how you feel. Especially if you found your breathing somewhat tight and constricted just now, as most of us do, let me guide you to say a statement describing your present condition, and then say an opposite positive one-liner to free that condition:

Again, tune into your breathing ... notice how it feels as you inhale and exhale ... now say silently to yourself on your next exhale: "My breathing feels tight and shallow."

Now, to purposefully let go of that habitual tense condition, say this positive one-liner to yourself several times on your next few exhales, "I set my breathing free." Notice how saying these positive words shifts your breathing in that freer direction ...

This is an effective psychological technique you can use whenever you feel or think something negative or limiting, or you feel bad or weak. In later chapters, we'll use this technique in depth to clear your mindset of outdated beliefs and rules and teach your subconscious mind new one-liners that will brighten and empower your life.

Now let's apply this de-beliefing process to what you've just been learning about the quantum mindset. First state the old Newtonian belief about who you are physically, and then come up with a Quantum One-Liner to re-program your subconscious mind with the model and liberating vision of who you are.

Let's take the quantum fact that you are not only mass, heavy, solid, and unchanging; instead, you are mostly very light, and constantly-changing energy:

To make this process most effective, first tune into your breathing ... and your whole-body presence here in this emerging moment ... rather than ignoring or running away from your current inner condition, go ahead and accept all your feelings right now ... again say silently to yourself on your next few exhales: "I set my breathing free."

As you continue breathing freely, tune into the actual presence of your physical body ... be aware of your feet ... your legs ... your pelvis ... your torso ... your throat and tongue ... your jaw and lips ... your nose and eyes ... your head ... open up to experience all those billions of cells and atoms and electrons and quarks that are your body ...

Now, first say the Newtonian one-liner belief that you were probably taught as a child: "I am a solid physical body." Feel from the inside-out how this ingrained belief tends to generate the actual bodily sensation of being solid, heavy, massive, and mortal.

Now, in harmony with the new quantum mindset that you're encouraging, say this Quantum One-Liner silently to yourself several times on your next exhales: "I am a mysterious energetic presence."

After saying this one-liner to yourself several times, just sit quietly, breathe freely, and let these expansive words continue to resonate throughout your being for a few moments—"I am a mysterious energetic presence." ... allow the words to generate a deep inner experience to match the words.

This process is a powerful way to consciously awaken your new quantum mindset!

Part 2

Making Quantum Changes In Your Life

Chapter 6:

LOVING ENTANGLEMENT

We've seen thus far in our discussion how everything (and I mean every tiny particle in the whole universe) is somehow energetically in relationship with every other particle in the universe. This might be the greatest mindset leap of all. When we fully embrace this new scientific theory, we can let go of feeling isolated and alone in the universe. As we'll see, this solitary disengaged belief is just that—an assumption we've been taught, that even our senses teach us, but which in the larger and also the microscopic and sub-atomic levels of reality simply isn't valid.

Some people seem to cruise through life relatively easily and happily, with things they want coming to them effortlessly, and sometimes almost as if by magic. They somehow attract to them what they want, but as we're seeing, this isn't a slight-of-hand magic trick. Instead, they've been lucky enough to inherit a set of genes and a world view that enable them to feel continually engaged with the world around them.

By living within the now-verified quantum belief, they're able to stay open to participating enthusiastically in a world of abundance and fulfilling relationships. If you're one of these people, count your blessings! If not, or if less so, in this second part of this book we're going to explore how to consciously let go of old attitudes and rules and embrace a vastly-wider world view where you too can partake in all of your potential blessings.

Here's a story I like very much that expresses in a metaphor how our beliefs and attitudes and mindsets determine how our lives will unfold, and what we need to do to optimize the flow of good events that might come naturally to us:

A grumpy old man dies feeling unfulfilled, unhappy, and miserable. When he arrives in heaven, he sees a big beautiful golden box wrapped up with his name on it. He asks, "What's in the box?"

His guide hesitates, then says, "Well, normally we don't tell people what's in the golden box, but since you asked, it is your box, and you can open it."

The old man opens it up and inside he sees all the items, experiences, relationships, and opportunities he didn't accept and allow into his life. These are all things he wanted, all the gifts from God and the Universe that, because of his ingrained negative mindset, he rejected, avoided, destroyed, denied, or discarded.

At that moment (and too late), he realized in life he'd been offered everything he could possibly want or imagine, and the only reason he didn't receive and enjoy them was because he did not accept them. He was so closed-minded, defensive, and negative that not only could he not accept them, but he couldn't even acknowledge their availability.

This story expresses how our habitual assumptions and limiting rules blind us to possibility. We go around with mental and emotional blinders that sorely limit our view of what life has to offer. So often we're offered everything we need to be happy in life, and all we have to do is see the opportunity and gracefully and thankfully accept and enjoy our gifts.

The most common subconscious sabotage is the belief, "I do not deserve wonderful things in life because I'm not good enough or smart enough". I love the golden box story. It says we're all given everything we want in life, and it's our choice to accept or decline these things based on our self-worth, beliefs, and internal values. Choice is the greatest gift of all, yet it can be the biggest curse as well. We can choose to be happy and create a happy life, or we can focus on the negative and attract more negativity.

In the quantum vision of probability and possibility, each choice and decision we make influences whether we're moving toward or away from what we would like to attain. We *can* have the monetary security and satisfying relationships we want. No one is stopping us from making the choices that lead to fulfillment, but to see our opportunity, we must believe in our higher potential. We must transform our attitudes so we can perceive an opportunity, respond with focused attention, mutual attraction, and active engagement, and participate in the dynamic dance that will bring us together with what we want.

Believe ... accept ... act ... receive!

However, if you feel you don't deserve a beautiful, happy, wonderful career or marriage, for instance, you predictably won't receive it even though it's always available. In the next few chapters, we'll deal with any old tendencies you might still be packing that might make you reflexively negative and operating out of limiting self-defeating rules and contractions. Then hopefully at metaphoric levels, when you get to heaven, your box will be empty, and your life will have been full!

We've already seen that manifestation and attraction (and literally everything else) run on a mysterious energy source that scientists can't as of yet see, much less know how it originates. Many ancient traditions spoke about and worked directly with this life force, calling it *aka* (Egyptian), *qi* (Chinese), or *prana* (Hindu). We know psychologically when we let fear, denial, rejection, and disbelief rule our life decisions, we disconnect ourselves from the life energy flowing through us.

Manifestation is all about receiving and giving energy; it's having the faith and the courage to perceive and accept all the energy that's being given to us, to use it to create what we want, and to freely give energy back to the world by letting it flow through us to others, the flow of love, blessings, riches, health, and true fulfillment. This openness to receive and give equally is ultimately what makes the difference between having and not having.

FREEDOM OF CHOICE

We tend to become set in our ways as we mature, grow up, and grow older. Our inherited mindset is full of rules and apprehensions that try to ensure that we won't get hurt or suffer disaster—and this is good! We carry hundreds of one-liner unconscious guardrails automatically keeping us from believing or doing things that might harm or even kill us.

We need a healthy balance between staying safe and limited, and entertaining new ideas and possible actions that can expand our lives and bring us more fulfillment. Even though our unconscious regulations and ingrained traditional limitations keep us safe, we need to stay conscious of new choices we can make to override and transcend old inhibitions.

Tapping into your conscious mind's power to override old beliefs is key to mastering the Quantum Manifestation process we're exploring here. Whether you act on them or not, you always have the freedom to look in new directions, think new thoughts, and entertain new possibilities in your creative mind. As we're seeing, the brain has the amazing capacity to bring together your sensory experiences, your logical ideas, your vast stores of memories and past experience, your almost-mystic power of imagination, and your even-deeper intuitive flashes that integrate everything into a new vision of what's possible.

Notice it all begins with choosing where you're going to focus your power of attention, each new moment of each new day. You make thousands of discrete choices every day, but how many of them are conscious choices, and how many are made without any conscious decision at all, deep in your subconscious mind?

I encourage you to allow your subconscious mind to continue to make all the automatic decisions that run and regulate your body quite effectively without conscious attention. I also encourage you to become more conscious of the key choices you make which perhaps could be changed to alter your life flow in new desired directions.

Although many people fail to use it, your primary power that no one can take away from you is the freedom to focus your attention anywhere you want, each and every moment of your life. You alone decide continuously what you think about and experience. Unfortunately, you probably developed ingrained attitudes and negative thought patterns in childhood that still tend to dominate your thoughts and choices in adulthood.

Choosing to pay more attention to how you use your power of attention is a major first step in transforming your mindset. When you focus your attention on anything, external or internal, you're directly aiming your personal power in that direction. In the process, you're energetically impacting whatever you focus on, as quantum theory explains. This seems so obvious, but it's so often forgotten and ignored.

Now we come to the question of why you aim your focus here, rather than there: Who is in charge of your mind? What motivation decides where you spend your most powerful charge of energy?

This is one of the core questions that has transformative power in and of itself. Self-awareness is observing how you determine where you're going to focus your attention next:

If I say to you, focus your attention on your breathing right now, you will almost automatically do this, unless you resist the suggestion. If I say, now focus on someone you love, notice what happens to your attention … If I say, focus on what's most important to do next in your life, again just notice how your mind's attention shifts, and toward what?

There, that's your omnipotent power of attention in action! This is your primary manifestation tool, so throughout the rest of this book we'll be regularly exercising your power to take responsibility for where you choose to focus your attention. You know from the first part of this book that you are pure energy. You pack quite a charge! Only you can direct your power of attention so that you focus on what's most important!

This choice-power will become super important when you start to choose what you really want to manifest next in life. If you believe you have limited choices and you already know what they are, you'll just continue to putt along doing the same old things. But if you expand your world view, entertain novel ideas, and imagine new developments, you'll set yourself free. As every wise parent says to their offspring, "Choose well, my friend!"

CHOOSING AUTHENTICITY

One of the primary choices you'll need to make more often to empower your life is the decision to be honest and authentic, rather than choosing out of default to put on a phony face and pretend to be the way you think people want you to be. This is crucial. When you're phony, you lose your personal power. You must be real to accomplish real things.

Honesty is a core human virtue in all cultures because being dishonest disconnects you from the reality matrix in which you live. This chapter is about entanglement and engagement, and if you're not being authentic, if you're not grounded in your personal core of being, there's no way anyone or anything can realistically engage with you.

Anything you want to manifest will require that you be real, that you be you, be honest and true. But we're so conditioned to fake it! The socialization process that all children must go through before the age of six is riddled with situations where we had to lie, fake it, and deny our gut feelings to avoid punishment and to "fit in" to a social group or family circle. We were punished for doing "bad things" even when we were toddlers. Long before we could develop a sense of ethics and morality, of right and wrong, we were forced to be dishonest so we would not get punished for doing what we wanted to do.

This is how life is in every family and community on the planet. We often had to lie to avoid punishment, then when we entered school, we found ourselves getting punished for telling a lie, and of course we felt confused! Now as adults, we must deal with our ingrained habits of pretending, lying, and faking it just to get by.

In a way, socialization can serve you as you develop a unique expanded mindset and world view that will truly be "yours" and not just whatever sense of ego-presence your parents and community wanted you to have. I'm not blaming your parents and community; after all, socialization is essential for community. We do need a set of accepted social rules to live by if we're to get along harmoniously as a group. However, we also have the inherent right to grow up and make our own honest choices, rather than being dominated by put-on social rules.

To be authentic means being real, being true, speaking honestly, and being trustworthy and genuine. It means you're not fooling yourself or others, and you don't have a hidden agenda. It means honestly aligning your energy field with your surrounding matrix of quantum forces, so you become empowered to tap into the higher manifestation powers.

Consider how you feel deep-down about this whole notion of choosing to expose yourself, to reveal who you really are ... notice how your breathing is affected by this shift of focus toward your own true self ... do you feel you're usually quite genuine and up-front and honest with other people, or do you tend to put on a phony social expression when relating ... do you often feel one way and act another? Do you say you feel just fine when in fact you feel lousy inside?

Consider if you're honest with your own self ... do you have parts of your personality that you habitually bury or reject or judge as unacceptable?

Now imagine how you'd feel in your breathing and your heart right now if you choose to be your true self, not hiding or denying anything at all about who you really are ... how does it feel to be just plain old honestly-genuine you?

Making this choice to be authentic over and over again, day in and day out, is essential if you want to let go of inauthentic reflex habits, and begin to strengthen your true power, your genuine presence in the world. Here's a list of subconscious one-liners that most of us are packing, which seriously limit personal power and your ability to establish meaningful relationships. Read the list, saying each statement as you exhale and notice which ones resonate with how you usually behave:

"I put on a happy face when I'm sad."
"I pretend I don't care even though I do."
"I act bright when actually I'm feeling low."
"I often fake being smart when I feel dumb."
"I act like I'm confident when I'm anxious."
"I say I feel great when I feel like hell."
"I act like I know when I don't."
"I pretend everything's okay when it's not."
"I fake friendship when I feel distant."
"I act like I'm present when I'm really far away."
"I put on a welcome smile when I feel shut off."
"I act like I'm interested when I'm not."

PLAYING THE ODDS

At subatomic levels, nothing is certain. We can't even predict where a particle will be at any given moment. The universe runs on probabilities, not certainties, and this means we always have a potential opportunity to use the power of conscious choice to alter our predicted course. We always have multiple possibilities presented to us until we make our choice.

Before we commit to a choice and observe our selection manifesting in regular physical reality, everything exists in our minds as a probability wave. In the quantum view, we can't perfectly and reliably predict anything in life. Right until it manifests, a situation can turn out a number of different possible ways. Depending on how and when we focus on an emerging life situation, a number of quite different outcomes might occur as waveform potential becomes particle reality.

For several hundred years, traditional science has been informing us that we live in a predictable, deterministic universe where there is no choice; free will had become an illusion, a figment of the imagination. But quantum science has blown this deterministic world view to bits and opened up vast new realms of possibility. Anything that we can imagine is considered possible.

However, you must learn to believe fully in this new quantum vision of reality if you want to activate your manifestation potential. Remember that beliefs are defined as things we deeply hope to be true, but have no proof of them. Beliefs are probability theories; we gather as much information as we can about something's structure and behavior, so we can make a good guess about that thing's nature. Ultimately, it's just a theory.

People believed the earth was flat and you'd fall off the edge into the abyss if you went too far out to sea. Then Copernicus and Galileo observed with new telescopes that the planet is spherical. The Catholic Church condemned Galileo as a heretic for challenging the Church's belief that the earth was the center of the universe. But explorers took faith in the science rather than the theology of the times and sailed off trusting the new belief the earth was round. The old belief was discarded as more evidence proved it wrong.

False or outdated beliefs that go unchallenged remain operational until someone dares to test the belief to see what's true. If you don't believe you can stand up, you won't even try. If you don't believe that manifesting what you want in life is possible and therefore don't try, your old belief wins.

It becomes crucial to learn how to bring all your one-liner childhood beliefs and assumptions up into your conscious observation, so you can shed the light of adult reason and seasoned intuition on those one-liners. IF they hold up and are valid beliefs to live the rest of your life by, great—keep them! But if you can see a belief no longer serves you and isn't valid for your situation, you're free to put a new belief in the place of that old one and liberate yourself.

If there's only a one-in-ten chance you can win at something, do you take the chance? In a materialist world view, perhaps not—it's too risky. But in the quantum world view which includes both things seen and unseen and believes in them both, you can hedge your bet considerably by learning how to influence certain invisible variables to your advantage.

Even with just a one-in-ten chance of attaining your goal, quantum physics tells you that each time you make a choice that aims your power of intent in a particular direction in the hope of a particular outcome, your choice will move you a bit in that direction. The next choice you make to go further in that direction will have a higher probability of success than before. If you have faith and make 10% probability changes in a direction toward your ultimate goal, you'll probably end up getting it!

Consider this: in any manifestation process, you first envision a desired outcome, and then you make your first tentative movement toward manifestation. A process is a step-by-step affair. Each time you make a step, you can pause and review the progress, and then perhaps choose a slightly different tack toward your goal.

Whenever you pause and choose to act in that desired direction, your situation will evolve and become more realistic. You keep nudging your vision and fine-tuning it as you choose more specifically how to make your dream come true. Each step of the way, you can advance your understanding and intent as your initial dreamy imagination evolves into final physical expression.

There's something else happening in this basic process of manifestation. Quantum mechanics has revealed what intuitive folk have sensed and worked with for countless generations: the law of attraction is never a one-way affair. Both parties in any attraction drama are active participants. You're never on your own putting out an attractive force to bring something you want toward you. That other particle, person, situation, or thing is also mysteriously pulling you toward it. You've become energetically entangled with each other.

Let's look deeper into entanglement as it happens at subatomic levels, and then reflect on how this applies to the way our minds work within these quantum-attraction laws to affect the outside world.

MUTUAL ATTRACTION

Even back in Einstein's time, the notion of particles becoming entangled with each other began to appear in mathematical equations. Einstein found the whole idea of entangled particles at a great distance somehow communicating with each other—and doing this instantaneously—to be "spooky," yet quantum entanglement has persisted and is now considered a fact in quantum mechanics.

There are still giant unknowns regarding the scientific dynamics of particle entanglement, but with the little that is known we can see that fundamental truths are being revealed. For instance, when two particles are engaged with each other through various levels of encounter, a connection, a bond, a relationship is formed that endures. If you change one particle, the other changes too. They become entangled by some force that remains entirely unknown, but they are entangled!

Particle entanglement is quite common, it's not rare at all. Any two particles that interact with each other will probably acquire at least partial entanglement. For instance, all atoms have electrons swirling around their nucleus, and all electrons in an atom naturally become entangled with each other, and with the atom's neutrons and protons.

Any time there's an interaction, there's an opportunity for particles to exchange information about their internal states and thus get into an "entangled" compound state. It would be very unusual to find a particle that's not at least somewhat entangled with other particles. In larger physical objects, from your cell phone to your body to the whole planet, most if not all the atoms will be partially entangled with each other to some extent.

It's becoming more accepted, at least in theory, that entanglement is not dependent on space and time. Mathematical proofs document that you can move one of two entangled particles to the other side of the universe and the entanglement continues unabated. As quantum researcher Anton Zeilinger puts it: "… on the level of measurements of properties of members of an entangled ensemble, quantum physics is oblivious to space and time."

Even though there are still unknowns in this story, we can see the general direction that science is moving in—and it's not new. There is a space-time continuum within which all physical matter exists. Einstein blasted Newtonian geometry by introducing time as the fourth dimension constituting physical reality. It is probable there are one or more additional dimensions that aren't bound within the space-time continuum.

We already learned in string theory there must be additional dimensions beyond Einstein's four to make the math work. Thousands of years ago, Vedic and Taoist philosophers came to the same conclusion. Along with physical matter, consciousness is quite real, yet it doesn't function within the laws of physical reality. As they say, "There's more to life than meets the eye."

We should be careful not to read too much into the quantum phenomenon of particle entanglement when seeking parallels in our personal lives. However, there's *something* happening at subtle subatomic levels that draws us to particular people and brings us together with things and situations our conscious and unconscious minds are focusing on. This is a growing belief in quantum circles, and it opens up vast new possibilities.

What's important for us here is the evidence that all particles interact, that there's a force or communication dynamic of some kind that exists naturally throughout the universe. Because the human body consists of massive bundles of interacting and bonded particles, these forces and laws must somehow apply to how we interact in our everyday lives.

As you reflect on this new vision, let's bring the phenomenon of consciousness into the middle of our discussions of quantum entanglement, quarkian attraction, electromagnetic bonding, and human powers of manifestation.

MENTAL ALIGNMENT

Imagine someone is living alone and feeling a very powerful yearning to share their life with someone. By nature, this person has a particular overall resonance that is the energetic harmonic expression of the totality of who this person is—her emotions, her thoughts, her health, her sexual charge, her inner experience, and her outer activities.

Now imagine there's another person living alone on the other side of the world. Let's make it a majority-relationship where this other person is a man. He has his own energetic resonance that expresses the totality of who he is at any given moment. He's there night after night, ruminating and mediating, yearning and imagining, broadcasting his yearning out into the universe while she's doing the same.

I have a friend, Tom, who was in this situation. He was working on the family ranch in Idaho while taking a break from over-work in a nearby city and recovering from a divorce. When he paused for half an hour each evening and did his daily meditation, he started to feel a deep yearning in his heart and mind that became quite specific. Each time he tuned into his romantic yearnings, he felt that there was someone "out there" also focusing intimately on him.

Tom's energetic yearning went on for months. Then, seemingly out of the blue one night, the ranch phone rang with a fellow in Europe inviting Tom to come do some seminars on a self-help book he'd recently published. The trip seemed preposterous to Tom. He had no interest in traveling, so he said no (his choice).

The next evening, the phone rang again, and it was the same German man asking him to please come. Tom again chose to say no. On the third night when the phone rang, Tom felt a leap in his heart that pushed him spontaneously to say yes, and three weeks later he was flying off to Europe.

You can probably guess the rest of this true story. Tom stayed at a guest house during his stay in Switzerland. His flight home was cancelled due to weather and he had to stay there for four days with nothing to do. For two days he felt strangely at peace, utterly content, and deeply tuned into someone's presence in his heart. Then on the third night, the front door opened, and a mystery woman walked in to check on the cats, assuming the American had already left. And pow! They looked into each other's eyes, felt each other's resonant presence, and fell in love at first sight.

The phenomenon of "love at first sight" is universal in all cultures of the world. It's considered a mystic occurrence beyond all scientific explanation. But we're coming to understand what might be happening here, if we are open to the possibility that there are levels of consciousness which transcend time and space, and subtleties of resonant entanglement that link two hearts and minds together—and bring them step-by-step toward each other until they meet.

I'm not expecting you to believe this. I'm just asking you to entertain the possibility that this might be true. A basic premise of this book is that the laws of quantum physics and the workings of the human mind must be one and the same. We have loads of divergent theories, but we're all living within the same reality. If "as above, so below" is true, then the universal force, intent, and intelligence that originally brought the first electron into bonded engagement with a lone proton to form the first hydrogen atom, this same basic dynamic could be what draws two people of similar resonance together to form their own bonded unit.

If this metaphor is valid, at the quantum level and beyond, operating outside of or beyond the space-time continuum, we can broadcast our need, our yearning, our intent, and let the entire universe know what we want. The overall model is clear and plausible: we send out our clear intent and charged desire, and the intelligence of the universe receives this broadcast and responds in specific ways.

Why not? In a universe where everything is connected by dark matter and dark energy and who knows what else, the intelligence and compassion of the universe could instantly align physical reality so that, in a seeming coincidence, we encounter who and what we seek.

In this world view, we do best when we expand the notion of the mind to include not just the brain but the whole body, packing its unique charge and presence. Humans naturally have a large set of needs, desires, yearnings, and requirements in order to feel safe, happy, and fulfilled, and these are felt in the whole body, not just the mind.

Perhaps the invisible force that brings two similar people together—the power of love—isn't just an ephemeral romantic illusion. Maybe it's a real force just like other real forces that scientists know are there because of how they impact the world, even if they can't be detected. This is the "spooky" realm where belief and fact, imagination and proof, border each other.

Remember what we learned about the strong nuclear force that mysteriously holds the proton together with the neutron? Remember the electromagnetic force that holds the proton together with the electron? What if there's also a natural force that brings two hearts together? From the logic of quantum science, I think it's plausible to believe there's a natural force called the power of attraction that brings and holds two people together.

> *This feels like a good place to pause and let you reflect. In exploring the manifestation process of "believe … imagine … act … succeed," we've spent over a hundred pages challenging you to expand your mindset so you can risk taking a quantum leap into believing in at least the possibility of an invisible attractive force in your life. If you look to your breathing … your whole-body feelings … your heart … and your inner wisdom, what do you find right now? Do you reject this belief? Do you embrace it? Could you tentatively entertain the notion and see how the belief feels as we move forward?*

THE ATTRACTION PROCESS

The following is a beginning outline of the Quantum Manifestation Process being presented in this book. As we'll explore, you can apply this method to anything you want: a new relationship, a new computer, a new outfit, a new job, or a new invention. The choice is yours!

To begin any act of manifestation, first you need to feel a yearning, a need, a passion, a goal, or a dream with a definite energetic charge associated with it—a whole-body feeling of need or desire. Perhaps you'll feel this charge as an emotion. Maybe you'll feel your mind buzzing with a budding idea. Memories may be involved, and surely imagination will be. Often there will be something in your environment, on television perhaps, or something a friend said, which will stimulate you to focus your attention in the direction of something you want to bring into your life.

Then comes the budding of an inner vision, a visualization, a clearer sense or idea of your intent. This inner vision will need to be imaginable; it must be something your mind can see, feel, hope for, and believe. It also needs to stimulate the development of an emerging objective, so you can begin to imagine the experience of actually attaining your goal.

Holding this vision of future completion in your mind will generate a beacon that gives you a sense of where you're heading. As you envision the desired end result, you'll help create that result. At this level, I suspect you will tap into the realm where past, present and future operate together, rather than in a linear relationship. This is called retrocausality, where the future somehow influences your present moment.

When you know in your mind and heart what you want and can see it in your mind's eye, you'll find that you begin to naturally broadcast your desires and intents outward, in the direction you expect to receive what you seek. This might start as very fuzzy and uncertain, but the more you believe in manifesting your dream, the clearer that dream will become.

You might also find when you fall asleep at night, you're effortlessly dreaming about your goal. This can provide a powerful boost of unconscious forces which are aligning with your conscious intent. You might also start having lucid dreams during the daytime that somehow point you in the desired direction. Your subconscious is now helping you to make your dream feel more and more real to you.

At some point, the "innovative how-to" function of your mind will start to come up with concrete ideas and action-steps that move you forward toward materializing your goal. What you then put into action doesn't have to be anything in particular; it just needs to be definite action that you take toward your goal, demonstrating to the universe what you want to express in the world.

To guarantee your success, you'll need to apply the power of your mind to continually visualize your goal as if it has already happened, as if this goal is already a reality, a certainty with no chance of it not happening. This power of belief, as we've seen, is invisible and unde-tected by science. It's one of those so-called imaginary elements like dark energy and the "i" in math. It's mysterious and unseen, but when you take it away the manifestation process stops. You have to *believe* in your manifestation power, or you simply don't have that power.

SELF SABOTAGE

Along the way to manifestation, notice if you have negative issues with your goal, if somehow you're feeling uncomfortable or anxious or out of harmony with your intent. If you feel uncomfortable as you make certain actions toward manifestation, those actions should be reviewed. Either your goal needs to be altered, or some old fear-based beliefs are reacting to your plans.

This is an important time to pause and listen to your inner feelings, uncertainties, and worries. Set aside some time to retreat and reflect. Allow unconscious questions or hindrances to rise up into clarity, so you gain deeper insight into your process. This will allow you to fine-tune your trajectory.

When you pause and notice how you're feeling, be sure to be honest with yourself. Often, you'll find old beliefs are trying to derail your enthusiastic flow. You may experience self-doubt and almost give up your quest, just because negative emotions provoked by old disbelieving one-liners are dragging you down.

Fearful habits and old beliefs will rise up often to resist the attainment of your desire. In the following chapter, we'll explore effective ways to work *with* your unconscious mind rather than fighting against it. Make sure your early beliefs are still in alignment with where you want to go now. You have to believe in yourself and your vision so fully that it registers with your subconscious. Once your subconscious believes, it will work faster than you ever could to make your vision come true.

Devote time to focus on training your subconscious mind to resonate in harmony with your conscious vision. Perhaps you'll work with realistic confirmation-statements. You might record a specific, short affirmation of your current intent, and listen to it just before you go to sleep each night to fix the vision more firmly in your subconscious mind. Use whatever method that works best for you. Your subconscious mind sets your intention and your goals. Your conscious mind envisions and organizes intent, but it's the subconscious mind that goes to work and provides the special energy needed for quantum manifestation.

EXPECT MIRACLES

In the classic Newtonian world view, there was simply no room for miracles. Materialist scientists held vigorously to the belief that everything would somehow be explained away within the scope of physical logic and mechanics. The quantum revolution has dispelled the rather egotistical presumption that human brains can comprehend the dynamics of the universe and beyond.

In fact, physicist Dr. Sean Carroll recently noted the following in the *New York Times* article, "Even Physicist Don't Understand Quantum

Mechanics": "Quantum mechanics is the most fundamental theory we have, sitting squarely at the center of every serious attempt to formulate deep laws of nature. If nobody understands quantum mechanics, nobody understands the universe. But as the physicist and Nobel laureate Richard Feynman observed, 'I think I can safely say that nobody really understands quantum mechanics.'"

"If you want to find the secrets of the universe, think in terms of energy, frequency and vibration."
—Nikola Tesla

In my life, the more I apply quantum logic to manifesting what I personally want and need in my life, the more I become a full believer in this mysterious dynamic. I'm having great success in attracting into my life what I ask for, and sometimes it *does* seem quite magical, almost miraculous. I simply recognize a need in my life and then focus on fulfilling it, using potent attraction meditations, and the need is satisfied.

I often say to myself, *This is impossible. It can't be this easy. How can I accept this?* But it's true. All you really need to do is think the right way, and your life will become more fulfilled. Train yourself, practice the process, and you will get results. If you're not getting results, you're probably not practicing. Curiously, it's hard to accept that getting what you need in life can be so easy, and that's where most people get stuck. If you believe it must be hard, it will be.

Experiment with and explore your own manifestation powers, and you'll come to the realization that yes, it really is that easy. It's easy because you're putting aside beliefs that disempower you; instead, you're aligning your personal intent with the higher quantum intent.

TIP: Start with smaller manifestations that are more believable and can easily be achieved and measured, then work your way up from there.

When you accomplish this alignment of personal with universal, you can often experience what others might see as almost-miraculous manifestations. Conversely, when you start to sabotage yourself by refusing to believe manifestation can be easy, you eliminate the ease of that possibility. If you don't believe in something, you'll broadcast your doubt, and the power of that doubt will make it nearly impossible for your dream to come true.

I'm not talking about believing in any vague pipedream or made-up fantasy about miracles happening. Just the opposite. I'm talking about aligning with the core quantum principles that we now realize run the whole universe. All you need to do is expand your understanding of life to include the quantum dimension where everything is a probability and not an inevitable outcome. Then, with your personal imagination aligned and resonating with universal probability, you can indeed generate what seem to be miracles.

At the quantum level, your positive thoughts are generating a non-physical broadcast wave coded with your intent, that flows outward from your center. This broadcast of your intent impacts quantum realms of reality as the universe responds to your creative intent, and the seemingly miraculous happens. The truth is you're tapping into the *probability* that your wish can come true, and the probability of it happening is greatly increased by your thoughts interacting with quantum probability (wave alignment).

The following thought experiments bring up some deep questions regarding observation, perception, and quantum probabilities. There's a study called the double slit experiment that solidifies the notion that not only do we exist on earth, but that we are co-creators and active participants in the creation of this world just by being alive and conscious. Let me explain...

In quantum physics, everything is a wave and gives out a wave signal until it is observed by something. Electrons act as both particles and waves, as you will see in this next experiment.

In the double slit experiment, researchers shined single photons of light onto a board with two slits in it so the photons could pass through the two slits onto a larger screen behind it. The experimenters expected to see two rows of photons come out on the other side because they have two perfect slits that the photons can pass through. However, what they saw on the board was a wave-like interference pattern of dark and light areas instead of two rows. This perplexed them.

When they tried to observe this happening at the slits, the pattern on the screen changed from a scattered interference pattern to the pattern of two rows that they originally anticipated.

What they realized was, as soon as they put an observer in the equation, the equation changed from a wave into two slits. When they took away the observation, it became a wave pattern again. They performed this experiment many times and continued to experience the same outcome.

It was a wave until it was viewed by consciousness because apparently an individual's attention was enough energy to affect the outcome of the probability wave. Nobody knows why or how this happens, but the experimenters concluded that an individual's consciousness can affect the outside world without even doing anything, just by being an observer. Maybe this is why consciousness is so important and enigmatic to us. This could also create a debate that in order for something to be created, it needs to be observed by someone to make it part of our tangible world.

The Bible said we have been given more power than we realize; perhaps this is the power of co-creation, a power which we all possess effortlessly. Could it be that God, our Creator, in simply His or Her observation of the Universe, is the creation of what we see today? Could it be that being "...created in God's image," does not literally mean we look like Him, yet in God's image we were created because He observed and projected His consciousness onto this quadrant of the Universe? Or maybe we have been given His ability to observe and co-create through our power of awareness, consciousness, thought, and free will?

CURIOSITY KILLED THE CAT

Another example of this is Schrodinger's cat theory. This experiment states that if you put a cat in a bunker with poisonous gas, the gas has a 50% chance of killing the cat and a 50% chance of not killing the cat. So, until we look in the bunker, we don't know if the cat is dead or alive. When we do look in the bunker, the cat is either dead or alive 50% of the time.

The quantum interpretation of this experiment is that before we look in the bunker, the cat is in a superposition, both dead *and* alive, and only through our observation is nature forced to decide whether it's dead or alive. Basically, it states that nothing is "real" until someone observes it.

After reading the aforementioned studies, you may have a different opinion about the famous philosophical thought riddle, "If a tree falls in a forest and no one is around to hear it, does it make a sound?"

What if it's not just nature deciding once we observed? What if something else determined the outcome of Schrodinger's cat experiment besides our observation?

No one knows. However, maybe the outcome depends on the *type* of energy that's projected out from the observer. Maybe this is why it's so important to project positivity from your energy field into the world around you. We already know that the energy from our thoughts and words can affect the molecular structure of water, the most receptive element in the universe. We also know that a group of people's thoughts can impact random number generators from any distance.

With the double slit experiment and Schrodinger's cat theory in mind, imagine if the energy you unconsciously project from your thoughts and emotions due to your current world view directly affect the outcomes around you just by your observation. Positive energies and positive thinking tend to produce positive results via the law of attraction and reticular activating systems. If the energies are negative, everything will collapse more in that area. If the energies are positive, it will collapse more on the positive side. If this is the case, how important would it be to be more aware of the energy you give off and the mindset you have? It's Definitely something to consider.

The one thing we can conclude from these experiments is that we need an observer who projects awareness and consciousness in order to see all the probabilities and possibilities life offers. This could give new meaning to when Gautama Buddha was in solitude meditating under the bodhi tree and the demon Mara confronted him. Mara prevented Gautama Buddha's enlightenment because Gautama could not provide a witness to confirm his spiritual awakening. Buddha simply touched the sentient Earth with his right hand and the Earth itself responded with, "I am your witness." Mara and his minions then vanished, and Buddha was enlightened.

In the next chapter, we're shifting into a tighter focus on how quantum science perceives the human brain and the more-mysterious human mind, and also what are being called the transpersonal dimensions of consciousness. This is where individual and universal levels of awareness, compassion, integration, and empowerment align and work together.

QUANTUM FLIRTING

Let's end this chapter with an experiential exercise where you tune into the actual feeling of entanglement at the physical level of awareness. Whenever you focus your attention on something, someone, or some goal, by the dynamics of subatomic physics you're energetically engaging with that person, object, or goal. You're sending out your resonant presence at that moment, and you're also receiving the energetic presence of that object, person, or goal.

In a word, you're becoming entangled. This momentary engagement forms a bond that's often strong and lasting. I believe this is a valid metaphor both for how photons get entangled with each other, and also for how humans become entangled lovingly and otherwise.

Physicists are the first to admit they barely comprehend the strange subatomic phenomenon of entanglement, where two linked particles can be moved a thousand or even a million miles apart, yet when one changes, the other instantly changes as well. It does seem to be a miracle, especially the notion of shared communication occurring at a very great distance instantaneously.

But remember that consciousness itself doesn't seem to operate within the space-time continuum. The most advanced theory of physics, string theory, requires there be dimensions beyond the space-time continuum. Put those two thoughts together and it's quite possible for two entangled yet distanced particles (or people) to interact simultaneously with each other.

Carl Jung and Einstein talked together at length about this. They were pondering what Jung had coined as *synchronicity*, best defined as, "the simultaneous seemingly-meaningful occurrence of events which appear significantly related but have no discernible causal connection."

Jung, along with countless others before and after him, observed the mysterious phenomenon of synchronicity happening in his own life. Probably you have too—it's called "quantum flirting" in physics, and refers to the instantaneous sensing that two people or events are somehow linked in a higher-order, more meaningful way. Precognition, déjà vu, omens, and premonitions reflect this synchronous experience.

This instant energetic linking of object and subject seems to happen at subatomic levels whenever you aim your personal power of attention and your focused presence at anything at all. You might go for a walk and a particular tree suddenly grabs your attention. It's that grabbing experience we're talking about! It's when "subject and object" is put aside, replaced with two equal energetic presences tuning into each other at the same time.

Arnold Mindell, a highly-rational scientist, now believes whenever anything grabs your attention, what he calls "quantum flirting" is happening. In line with quantum mechanics, Mindell considers all matter on earth as conscious and intelligent in its own way and having the power to also focus on … you! Native traditions throughout the world, such as Taoism in China, have believed in this phenomenon since before the advent of written history.

Quantum flirting seems to happen all the time; we just don't usually pay any attention to it because our belief system says it's not possible. Every time something catches your attention, just before you pay attention and observe it, notice that it's doing the catching. In fact, if you start paying attention to this "catching attention" experience in your daily life, you'll realize that you can never tell who caught whose attention first.

It does seem simultaneous, but if consciousness doesn't function within the space-time continuum, it opens up an entire vast realm of possibility. It's exactly this level of new possibility that makes seemingly miraculous events occur. If you don't believe it can happen, of course you won't notice when it does. But if you don't *notice* something happening (because of your cultural belief blinders), does it mean it's not happening?

For instance, it seems possible, if we accept that dark energy and dark matter fill 95% of the universe (including right here on our planet), that invisible and non-physical energy and matter might form the matrix within which parallel dimensions abound. Entanglement and instant communication at a distance are phenomena that appear more plausible the more we open up to the full vision that quantum science is offering us.

Right now, do you feel connected at some non-physical energetic level with anyone? If you put aside beliefs and assumptions to the contrary, what do you actually feel? Tune into your breathing first ... then your whole-body presence right now ... tune into whatever feelings you find in your heart ...

If you feel in your heart a subtle but definite sense of loving entanglement with someone, are you going to say this engagement isn't real, or can you trust your heart and believe that entanglement is indeed quite real at some level of reality?

Now you can let that feeling go ... and choose to focus your attention inward toward whatever you might want to bring into your life in the near future to fulfill a personal need or desire ... perhaps it's a person you want to meet and engage with ... a new possession ... a group or situation ... a new job. See if you can sense a presence already out there—not only awaiting you, but equally flirting with your presence, and attracting your attention ...

Once you practice this primal attraction meditation a few times and begin to subtly sense the presence of what or whom you're seeking, you're already halfway to your manifestation goal. You've put away limiting beliefs and are exploring what happens when you entertain the expansive belief that this level of invisible energetic engagement is possible ... now you're ready to participate in this mysterious dimension and manifest your dream.

Chapter 7:

EVOLVING YOUR CORE RULES

In Chapter Six, we began exploring how your mind has been naturally programmed with a set of ingrained beliefs, rules, assumptions, prejudices, and fear-based limitations from a very early age. That's just how human society works; it programs its offspring with the one-liner rules and beliefs that maximize survival. In this chapter, we're going to look deeper into how your mind works and consider more closely what a quantum mindset entails. We'll uncover how the conscious and subconscious functions of your mind can be integrated to work together, rather than fighting each other.

We're living in a unique period of human history where technology and science are accelerating the pace of everything in life. Change is happening so quickly that most of us simply can't keep up. This means there's a time lag between when a change happens, and when your mindset gets updated to reflect the new situation.

In just the very recent past, no one knew how to use a cell phone, how to run a microwave, or how to jump on the internet and order a new pair of pants. If your great grandmother were brought back and introduced to all our new electronic tools and toys, it would all seem like magic. They would defy her senses and her reason, and, by extension, her entire world view. At first, she'd probably just disbelieve the new reality, and perhaps even fear it and try to make it go away and leave her alone. In fact, there are still some baby boomers and members of Generation X who reject some of the world's latest and greatest technologies because they still haven't adapted.

We're all going through a similar challenge. We can either reject the new information, proofs, and beliefs that science is throwing at us, and keep our world view intact, or we can manage our minds so that our world view expands, and we keep up with the speedy change around us. Our own minds are being challenged to change, but we're receiving very little help in making this change.

Obviously, I'm not suggesting you throw out your entire mindset that you've developed since birth (or even before birth), in order to keep pace. What I'm encouraging is a process in which you feed your mind with new key ideas, information, beliefs, and possibilities, so that outdated core assumptions of the past can be replaced with more scientifically accurate models and beliefs. This discussion has been doing just this—helping you to change your mind.

But what *is* this "thing" we call the mind? What have brain scientists learned recently that can enlighten this quantum discussion? I explained earlier that the word "quantum" isn't some magical incantation; it simply refers to the smallest possible piece of something. When we speak of the quantum mind, we're talking about a mind that dutifully adheres to the universal laws of quantum mechanics.

Deepak Chopra, an internationally-renowned physician, scientist, and philosopher, summarized this by saying that in quantum science, the universe requires an observer, and the human mind is capable of observation. In this light, he says that consciousness within the individual human mind acts as part of the universal mind. In alignment with quantum mechanics, consciousness is nonlocal like a quantum particle is nonlocal. Chopra furthermore believes that absolutely everything in the universe is entangled. Consciousness acts as a matrix, not a material thing; it functions as a non-physical realm of possibilities where anything is possible.

Of course, traditional brain researchers seeking a material explanation of consciousness disagree strongly with Chopra's beliefs. They admit the brain is composed of atoms, and atoms follow the laws of quantum physics, but as one of them stated, "The evidence is building up that we can explain everything interesting about the mind in terms of interactions of neurons. Is there any need to invoke quantum physics to explain cognition? I don't know of one, and I'd be amazed if one emerges."

I don't mean to denigrate the amazing traditional brain research being conducted, because it's leading to new breakthroughs in brain surgery and other fields. However, the current models of memory and cognition don't even attempt to include the new insights and dynamics of subatomic interactions. Neuroscientists admit they don't know what memory is, they only know what it does, and how it can be stimulated and modified. It's possible that for four thousand years, the ancient spiritual tradition of the Akashic Records has been right all along. Consciousness, and memory with it, might function outside space-time boundaries in a 5th dimension related to string theory and Carl Jung's collective unconscious. Time will tell …

Brain neurons are cells which are made of atoms, and atoms are made of quantum building blocks that we don't really understand. What we do understand is that the human brain, like all other giant organized groupings of atoms, must adhere to the laws of quantum mechanics, and that means that ultimately, the brain is operating within a nonlocal and nonphysical matrix. It consists of both matter and anti-matter, and, of course, dark matter and dark energy, not to mention the quarks and gluons and the rest.

At the level of quarks and gluons, we've seen that matter consists of nothing but pure intelligent resonant energy-bundles interacting with each other. Here we find the current materialist model of the brain confronting the mysterious experiential "thing" we call consciousness. Following in the tradition of Rupert Sheldrake, scientists are now trying to prove that tiny quantum-sized microtubules in the brain, somehow similar to gluons and quarks, function as the quantum foundation of consciousness.

Tubulin is a protein a thousand times smaller than a blood cell and is found in every cell in all plants and animals. It's a universal quantum presence in all of life. The proteins move chromosomes when cells divide, they enable communication at sub-cellular levels, and they form a cytoskeleton that supports the structure of all cells. They even form cilia that flutter outside a cell's membrane to generate movement. They're called the "transformers of biology" because they make things happen and are considered a core function of the life force.

Are tubulins a foundation of consciousness? Consider a tiny creature called the paramecium, which has no central nervous system. It has no brain or neurons, but somehow it can swim around, find food to eat, even find a mate, and know how to avoid danger. It can process information and seems to make choices as well. How does it perform this crude type of cognition? Well, its only internal structure are these strange universally-found microtubules which are the cytoskeleton of every paramecium.

Renowned scientists like Sir Roger Penrose have suggested these microtubules might be the quantum foundation for human memory, cognition, imagination, and just possibly, consciousness itself. Like quarks, microtubules are so tiny that even recent experimentation still hasn't shown us much about them. The research points in a general direction where our brains, our minds, our inner experience, and our consciousness all function at a quantum level that follows the emerging laws of quantum mechanics.

As you learn about this new model, perhaps the information is resonating within your microtubular level of consciousness. This new model and theory certainly won't be the last word on consciousness and how your brain works but understanding the model will expand your mindset in a meaningful direction.

EVERYTHING IS ALIVE!

Everything has some degree of consciousness, because everything is energy and energy is everything. We're all connected to quantum waves of energy and we're all feeling and co-creating the ripple effects of this energy through our observation.

It may be hard to believe that everything is alive, and everything has consciousness, but it's true. Everything from plants, animals, and even crystals (yes, rocks), are alive and hold some form of consciousness. You don't have to take my word for it; take the word of the smartest man in the world according to Albert Einstein. When Einstein was asked "How does it feel to be the smartest man in the world?"

He replied, "I don't know. You'll have to ask Nikola Tesla."

> *"In a crystal we have clear evidence of the existence of a formative life-principle, and though we cannot understand the life of a crystal it is nonetheless a living being."*
>
> —Nikola Tesla

Scientists conducted studies where they boiled eggs and they were getting emotional reactions from the eggs while being boiled. The same studies were done with plants.

In the 1960s, Cleve Backster, founder of the FBI's polygraph (lie detector machine), conducted a study on the plant kingdom. He took a plant and rigged it up to his polygraph machine to measure the electrical resistance within the plant, the same as he would a human being. To detect the vegetation's inner energy, he attached electrodes to the leaves and recorded the electrical impulses of the plant. This study monitored the energy in all areas of the plant (roots, leaves, soil, etc.), while Backster recorded the causes and effects when performing certain procedures such as watering the plant. Other more invasive testing proceeded, such as dipping the leaves in hot coffee and even burning the leaves of the plant with matches.

What happened next was so bizarre that it still leaves scientists puzzled today. The polygraph machine displayed an undeniable influx in electrical impulses when Backster burned the plant leaf. This proved to Backster that the plant indeed had some type of feeling that created an emotional response, the same as a human would when burned.

That said, just know that everything is alive, everything has consciousness, the universe and everything in it is constantly communicating with your subconscious at all times. Let's go deeper into what your subconscious mind is, and how you can jump right in and begin redesigning parts of it to your great benefit.

WHAT'S DOWN BELOW

We tend to throw around the word "subconscious" without quite knowing what traditionally the word has meant. We all have one, but by definition it's mostly invisible and perhaps even unknowable. The term was coined originally in the late 1800s in French, *Subconscient*, by Pierre Janet and then popularized by Sigmund Freud after 1993 (who later shifted to using the term "unconscious").

Carl Jung, who founded analytical psychology, soon used the term differently from Freud, but the general meaning referred to a powerful mental process that operates outside of consciousness. It was understood that subconscious thoughts, associations, memories, and impulses operate without conscious intent, control, or observation, and can often rise back into consciousness.

For Jung, the subconscious mind was extremely powerful, determining much of a person's conscious thoughts through archetypal images and shared cultural one-liners or societal rules. As mentioned before, Jung also coined a new term in 1916 called the "collective unconscious" ("the soul of humanity at large") which would prove to be a psychology-based forerunner of quantum physics.

The collective unconscious, a shared transpersonal level of consciousness, seemed utterly mystical a hundred years ago. The model is now making more sense scientifically because it so aptly describes a universe where consciousness exists everywhere, is nonlocal, and instantaneously connects individual minds, as mentioned earlier, Jung also coined the term "synchronicity" to explain how two individual minds are connected through the collective unconscious so that they are linked (entangled), can function as a unit, and share experiences together beyond space-time logic.

Let me ask again: When a thought flash into your mind, where did that thought come from? In Freud's perspective, it arose from the garbage bin of thoughts, feelings, and experiences you were afraid of and refused to accept in your conscious mind. You repressed them and stuffed them down out of sight. Jung saw this differently. He felt that along with the personal memories and imaginations you might have stuffed down out of sight, there are also a many positive aspects of the subconscious mind.

If you compare the one-liner beliefs and assumptions you carry in your mind, you'll find that most people in your society share the same sayings and rules. Many of the one-liners reflect simple experiential wisdom and logic: don't touch a hot stove, don't run or jump in the dark, don't provoke a bear, don't light your own tail on fire, don't tell lies, don't kill people, don't do this or that. Think of the Ten Commandments, for example. They just make good sense.

However, many of the one-liner rules and assumptions, as we've seen, reflect emotional constrictions your parents probably inherited from their own parents, and are not necessarily optimum rules for you nowadays. Specifically, there are one-liner assumptions and condemnations that came from uptight religious judgments and prejudicial tribal slurs that are unfair, demeaning, restrictive, inhibiting, and downright unjust.

Then there are the rules intended (often in a mean attitude) to control rowdy children and knock down budding geniuses so they wouldn't compete with parents, teachers, siblings, and other children. Kids are programmed with negative one-liners like, "I'm no good," and, "It's wrong to have fun," and "What I'm feeling is sinful," etc. These rules subtly influence their later adult lives.

Most of these core rules and regulations were picked up early in childhood, before reasoning could think the rule through and decide if it's a worthwhile belief or not. They lurk in the subconscious mind unseen most of the time, but they can provoke related emotions that suppress your freedom of expression. They also tend to inhibit play, exploration, free thinking, spontaneous relating, and the expansion of your mindset.

In the opposite direction of personal freedom, a lot of us still pack nasty one-liners deep in our subconscious minds like, "The experts know better than I do," or, "Who am I to question authority," or, "It's dangerous to step out of line." Notice this: all negative one-liners tend to be fear-based. Fear is the easiest way to control people.

Society is based on conformity, and in some ways, this is essential, but the opposite is equally true. People need to be free to do what they want, to think what they want, and to feel and imagine and dream what they want, even if society as a whole doesn't like it.

What we're seeking here is a wise balance. We need to examine our unspoken, unseen rules and attitudes, keep the ones that serve us, and throw out the rest. It's time to clean house, to clear space, and to lighten the load so we can fly! This process requires a bit of awareness, self-reflection, and mental work, but it's well worth it. You don't have to spend years in the process ; begin the habit of catching negative one-liners whenever they arise in your mind, and deal with each properly when it appears. Just remember that every trigger has the potential to make you bigger. Every time we become aware of an unconscious habit, we have the opportunity to grow.

Review your rules. Make sure you do not have any that get in your way. Like, "Rich people cannot be happy." This is not true at all, nor is "Money the root of all evil." Money is simply a tool for freedom that enhances the person you already are. If you're an angry jerk, you will continue to be angry and be more of a jerk if you have money. If you are a nice person and love to give to charity, you will continue to be nice to people and give even more to charity. Money doesn't make you; it just brings out who you already are on the inside because you have more resources to express yourself fully.

You'll be quite surprised (and thankful) to find that your subconscious mind, when approached honestly and in a friendly manner, is your very best friend. It's a problem only as long as you reject its subtle nudges and unexpected interjections. Once you begin relating positively with your vast resource of subconscious feelings, ideas, insights, and visions, you'll discover, as Carl Jung did, that your subconscious realms are a seemingly-infinite goldmine of great inspiration, energy, and guidance.

THE IMPLICATE ORDER ... WHOLENESS

We tend to think of the conscious mind and the unconscious mind as opposites, and in a way they are. We can only be aware of so much at once, and that's our conscious experience. Everything else is relegated to the unconscious. In a quantum mindset, both are vital and need to work together in harmony to generate our experience.

The subconscious mind remains mostly a mystery. Psychologists still don't understand what a thought is, and when they say that thoughts "emerge from the unconscious function of the mind" they aren't telling us much. What we do know is that ingrained attitudes and rules reside in the unconscious, and considerably influence our conscious thoughts.

Therefore, we must find a way to integrate our mindset's conscious and unconscious dimensions. We also must do our best to deal with how the subconscious mind can undermine our emerging quantum vision of reality. At the same time, we'll want to explore how we can tap the power of the subconscious mind to manifest at higher levels.

The subconscious mind functions in layers, with loads of secondary rules, attitudes, beliefs, and assumptions right under the conscious surface, and a few core beliefs and attitudinal structures deep underneath. In the same way physicists have dug deeper from cell to molecule to atom to quark to gluon to pure intelligent energy, psychologists have been digging ever since Freud, Jung, and Reich to get to the underlying quantum structure of the subconscious (or unconscious as Freud called it).

As mentioned before, they found that primary one-liners determine the class of beliefs and attitudes that run your mind. If you can transform those core beliefs and assumptions, you can shift all your secondary one-liners in one fell swoop. We're going to spend the rest of this chapter and into the next identifying and beginning to transform those core rules and attitudes, so you can progress swiftly.

Let's take one of the greatest attitudinal rules as an example—it needs to be a primary focus for stimulating a quantum leap in belief-transformation. I'm talking about the core foundation of your mental functioning where you either see the world through exclusive judgmental blinders, or you learn to shift into the powerful mindset of inclusivity. When you shift from exclusivity to inclusiveness as a core belief, you rapidly transform your life for the better.

You can run your mind on mental software that looks at the world around you and inside you as "It's either this or that"— and reject, judge, and repress half of your life. By holding onto this mindset, you'll stay stuck in denial and fear-based battles against everything that you don't habitually accept. This is how most of us have been programmed. It's safe but limiting, and often unfair.

Or you can take the quantum leap and make the choice to see the world as "It's both this and that"—and be welcomed into an entirely different world. You'll shift from exclusive rules to inclusive rules, and this changes everything. It's a genuine expansion of consciousness because you now include everything, and you accept reality as it is rather than denying much of it because it doesn't fit your old judgmental attitudes and tribal world view.

One of our great philosophers of the last generation was a British scholar named Alan Watts, who helped guide the sixties revolution in positive directions, and influenced millions of Boomer Americans. One of his primary contributions was his insistence that we strive to see the world as 200% rather than 100%. He encouraged his students to let go of "either-or" exclusive mentality where you have to reject something to accept its opposite. Instead, he taught how to see the world through inclusive "this-and-also-that" lenses.

When you see everything you encounter in this new 200% way, everything in your subconscious mind will respond and make the shift. As you develop the mental habit of being inclusive rather than exclusive, of staying open to both sides rather than being negative toward half of life, the root computer function of your mind will automatically begin to perceive each new situation in this inclusive light.

The ancient Taoist tradition in China, and later the Zen Buddhist tradition in Japan, fully embraced this inclusive vision of the whole. They developed a mental model that included both positive and negative within a larger reality. In the same way that opposites attract in science, negative and positive in the Zen understanding are integrated into a balanced 200% vision of life. The negative is 100% of its own full presence, and the positive is also 100% whole and complete. The quantum leap of inclusion brings both together into a larger inclusive reality.

Think of what we've learned about the atom for instance, because right here we can see this essential quantum truth. To have life in the universe, indeed to have a universe at all, the positive and the negative, the positively-charged proton and the negatively-charged electron, must function together in harmony, otherwise at the quantum level, the universe simply wouldn't work at all.

Everything we've discussed in this book so far comes together with this shift into accepting both sides equally, and unifying dualities into a higher-order integration. A number of basic human virtues since antiquity have endorsed this inclusive stance in life: tolerance, compassion, truth, and honesty all lead to an expanded mindset where all of God's creation, as they say, is honored and included within our mindset's inner acceptance circle.

So, I encourage you to take this quantum leap in your own mind, especially related to rules and attitudes and beliefs. No more either-or." Instead, shift to "this-and also that." In Zen terms this is the ultimate transcendent model, where negative and positive, yin and yang, female and male, and all the rest of the archetypal dualities are brought together into a larger sense of reality.

In essence, this represents the union of two 100% truths and conditions. Here's a visual representation:

In this world-famous Taoist/Zen drawing, the opposites of black and white remain separate, but they're enclosed in a membrane circle that holds both equally in balance, as one whole complete presence. This quantum leap into wholeness fully accepts the world of duality that we all live in, but rather than taking sides in a "right or wrong" mental attitude, we can vastly expand our mindset to perceive the unifying wholeness beyond the duality.

Niels Bohr, the quantum scientist we talked about earlier, used this yin-yang model to show how a particle can be a discrete physical object and also a mysterious wave function at the same time. In a similar way, the great American physicist David Bohm combined science and mysticism into an inclusive vision that he called the implicate order of undivided wholeness. He also developed the concept of "universal flux" in which everything moves together in an interconnected process, and is always in a state of "becoming."

Bohm taught with Einstein at Princeton. Together they furthered this new vision of the universe being in a state of undivided wholeness nothing is excluded, everything has its place, and all individual units are working in harmony within a larger integrated unity. To get a clear sense of what is meant by quantum wholeness and inclusivity, Bohm's writings are an excellent place to look!

Let's pause and give you a bit of time to reflect on what we've been exploring. How have you been responding to this notion of letting go of the old either-or mindset you were programmed with in your culture, and leaping into a quite different world view that's inclusive rather than exclusive?

Tune into your breathing and notice if you're feeling tight or expansive, as you consider entering into a truly transformative world view. As you stay aware of your feelings and your breathing, for ten breaths or so see what thoughts of your own might rise up from your unconscious realms …

SUBCONSCIOUS WISDOM

Most people consider Sigmund Freud to be the main historic figure regarding the unconscious mind, but he seems to have had his own blinders regarding the full nature of that unseen realm of the human mindset. He saw the unconscious mostly as the basement place where unacceptable repressed sexual impulses, immoral thoughts, imaginations, and memories went to fume and smolder and occasionally erupt as weird dreams and thoughts which upset a person's normal conscious life.

Freud assumed that most of the contents of the unconscious mind are unacceptable or disturbing, such as feelings of pain, anxiety, doubt, and conflict. This is certainly part of the story, and he was correct in pointing out that the unconscious is continually influencing our behavior and experience, even though we are mostly u~~~ these underlying influences.

Freud didn't seem to appreciate the remarkable *positive* influence of the unconscious mind. He was a remarkable thinker who contributed greatly to our understanding of human nature, but his approach to the subconscious was limited to his primary fixation (sex) and his times (sexually repressive).

His student, then colleague, then collegiate antagonist Carl Jung, whom we talked about earlier, saw things quite differently. Jung was more of a quantum thinker than Freud. He focused not on repressed thoughts but on inspired thoughts emerging from the unconscious realms of the mind. He trusted the more mysterious dimensions of consciousness, and spent his whole life exploring human potential rather than Freud's human repressions.

Jung was remarkably advanced, a true seer among psychologists, with an inclusive mindset that came close to expressing a quantum vision of reality. For instance, he believed that the human mind is a self-regulating system, like a living cell or body, that naturally does its best to maintain a healthy balance between opposing feelings and ideas, while also striving to mature into a highly conscious integrated organism or person. Here's a quote from Jung explaining exactly what he thought the unconscious is:

> Everything which I know but am not at the moment thinking; everything which I was once conscious of but have now forgotten; everything perceived by my senses, but not noted by my conscious mind; everything which, without paying attention to it, I feel, think, remember, want, and do; all the future things which are taking shape in me and will sometime come to consciousness; all this is the content of the unconscious... Besides these we must include all more or less

ressions of painful thought and feelings. I call
se contents the "personal unconscious."

nscious was extremely important because it held
ht need to access at any moment, not only to sur-
fulfill our highest potential. I find his mindset in
this regard wonderfully inspired with the vision from which quantum
science has emerged. In fact, as I mentioned earlier, he was of the same
generation as Einstein, and from Jung's perspective of the "collective
unconscious" he participated in that general quantum revolution that
swept the minds of great thinkers around the year 1900.

Here we come to another core belief or assumption that defines the
quantum mindset: the perception of our own unconscious mind as
either a negative ugly repository of repressed impulses (Freud) or an
inclusive and mostly positive goldmine of everything that makes us
who we are, and can help us become our full personal expression of
human potential.

*Consider this for a moment, as you stay aware of your breathing
and feelings: the thoughts that you think often seem to emerge out
of the blue, which means they come from your unconscious mind.
Do you trust your unconscious mind to be the generator of the
thoughts that come to mind? Are you feeling good about who you
are at unconscious levels, or are you distrustful of your own source
of inspiration?*

*I challenge you in the days and weeks to come to be more aware
of your thoughts as they arise into conscious awareness. Note that
most of your feelings seem to emerge out of unconscious thoughts
and impulses ... do you feel in harmony with this unseen generator of your feelings? Are you a victim of unconscious urges, or
blessed with them?*

TRUSTING YOUR INNER VOICE

In the next chapters, we will bring all this together into a concrete process for integrating your conscious intent with your unconscious desires, so your unconscious powers align with your conscious motivation. This higher-order integration of the two will provide you with the ability to manifest what you're needing.

Now we need to go one step deeper into your own unconscious realms, and consider your attitudes and beliefs related to your inner voice. Almost all cultures talk of the inner voice, and recently Western therapists have been working with what they call the "critical inner voice." This is a negative approach to where our thoughts originate and runs parallel to our discussion of "negative one-liners," thoughts that pop into our minds from the unconscious that are destructive, anti-self attitudes that discourage a person from acting in his or her own best interest. These are the nagging dialogues that try to control us out of fear and self-negation.

The critical inner voice is our lurking internal enemy, knocking down our self-esteem and confidence, undermining intimate relationships, and generating defeatist feelings of self-denial, self-criticism, and loss of motivation. Freud called this the super-ego which was mostly an internalized version of our parents and teachers, often at their worst.

We've talked quite a bit about how to deal with the negative one-liners that undermine your power to create and manifest. More to come on that in the following chapters. Right now, let's discuss the opposite: your true inner voice that expresses your higher yearnings and insights, goals and world view. When you begin trusting and listening to this inner voice of wisdom and encouragement, you can leap forward beyond most negative one-liners.

As teachers like Carl Jung and more recently Deepak Chopra insist, we all have an inner voice that we can trust—it's that sense of "knowingness" that transcends mental logic and gives us the feeling of certainty. Sometimes it might appear as an actual voice speaking to us from our heart, but usually it's more subtle, a hunch perhaps, or an image or impression. It feels like your whole body and being is gently nudging you in the right direction.

Sometimes it comes to you as a sudden inspiration, a realization, insight, or vision and your whole being responds positively to it. This is so important. This is where you tap into new ideas and visions that you will feel motivated to act on and materialize in your life. You're going to want to pay more attention to this inner insight source. It is the very beginning of anything you're going to bring into manifestation.

Traditionally there have been mixed beliefs about one's inner voice. Many religious theologies have scared people into believing that the devil will pretend to be God's voice in your head, putting evil ideas into your mind that will lead you astray. This negative belief might be disappearing in theological circles, but culturally it's still passed down generation after generation as a subtle one-liner saying, "Beware! Don't ever trust that voice!"

At the same time, many of us were taught that God or angels will sometimes speak to us, and that we should hear the voice and follow its guidance. Herein lies the confusion and conflict we've all felt when an impulse hits us out of the blue encouraging us to do something spontaneous. Should we trust that inner voice, or fight against it?

The solution is to get to know your inner voice. There are multiple voices that might speak to you, and it's your job to know when you're hearing your father's scolding voice, your own encouraging voice, or even God's voice. Which do you trust? If you start listening, you'll quickly learn to discern between critical inner voices from early childhood programming that put you down, and your own true inner voice you can trust.

Carl Jung coined the term "collective unconscious'" to indicate there's something else operating beyond your personal inner voice of wisdom. He believed (and so do I and a great many others) there's a vast, perhaps even infinite dimension beyond the spacetime continuum, in which a higher-order collective wisdom and intelligence, compassion and guidance is to be found.

Jung felt the integration of individual consciousness and the collective unconscious is our life goal, because this merger of personal and collective enables us to tap into a wisdom and truth that surpasses our personal capacity.

The quantum mindset is fully congruent with this model of reality, as we've seen. Again, I'm suggesting an idea, or perhaps you've already encountered it, that challenges you to take the leap into entertaining a possibility that can expand your mindset and boost your creative power.

If you don't believe in a higher-order source of inspiration and guidance, if you deny even the possibility, then you simply cut yourself out of ever tapping into that source. It seems wise to at least entertain the idea and see where it might lead you.

For instance, what does it mean "to listen to your heart?" Is your heart the consciousness-center where inspiration and inner guidance emerge? Do you encourage the inflow of wisdom by focusing on your heart?

As Krishnamurti often said, what's important is to participate in the inquiry, to be open to new insights, to allow your mindset the freedom to play with a new belief or model of reality, and see if the new reality fits better than the reality from your programming.

Let's end this chapter with some breathing room for you to just relax a moment ... let your mind calm down as you watch your breaths come and go without any effort ... begin to let your mind reflect on whether you yourself have experienced the inflow of an idea or insight or imagination or vision that seemed to just pop into your mind from beyond your personal consciousness ...

Now consider how you feel about flashes that come to you out of the blue ... do you trust them, and if so, do you want to encourage them to come more often?

Usually people tend to trust the inner voice when it feels like it emerges from deep within their own heart. You of course might have impulsive urges that come from repressed sexual desires and other sources ... but if you feel or hear guidance coming from your heart, do you trust that voice?

Let's end this experience with you turning your focus of loving attention to your own heart region ... breathe into whatever feelings you find there ... now see if there's anything your heart's wisdom wants to say to you right now ...

Chapter 8:

SUBCONSCIOUS POWER

Perhaps it seems a bit strange that we're spending so much time in a science-based book focusing on something as nebulous as our mind's gritty subconscious realms. In fact, it makes perfect sense, considering that we're also focusing on how to use insights from quantum physics to boost your power to manifest at higher levels.

It's now well-documented by research in neuroscience that only 5% of your cognitive activities such as making decisions, expressing emotions, moving physically, and reflecting, imagining, or planning are conscious, with the remaining 95% of your cognitive activity being generated in a non-conscious manner. When we consider non-cognition functions of the brain, such as running all the physiological processes of the body, that figure becomes even higher, with only around 1% being conscious.

"An enormous portion of cognitive activity is non-conscious," Dr. Emmanuel Donchin, director of the Laboratory for Cognitive Psychophysiology at the University of Illinois, has stated. "It could be 99 percent; we probably will never know precisely how much is outside conscious awareness."

Not only that, but all the unconscious mental functions of your brain run vastly faster than conscious cognition, occurring at the tremendous speed of just a few milliseconds, or about 400 billion bits of information per second, with impulses traveling at a speed of up to 100,000 mph. Compare that with your conscious mind which processes only about 2,000 bits of information per second at around 100-150 mph.

The subconscious mind mostly runs the show in terms of cognition and decision-making, not to mention all the auto-functions of the body. Furthermore, the subconscious mind is vastly more efficient than the conscious mind, and deeply influences everything you think and do at any moment. Clearly, you'll do well to harness the subconscious power of your mind, as we've been suggesting throughout this book.

The rather startling 5% statistic of the brain's cognitive function is the same as that other 5% figure we mentioned earlier, that only 5% of the universe's entire mass and energy is detectable with experimental methods, with dark energy and dark matter taking up the other 95%. Somehow our inner realms of awareness match the percentages of the outer realms of astrophysics.

Every decision you make, every emotion you feel, every thought you think, and every action you take are 95% driven by your unconscious function. This is especially true when you decide where to focus your power of attention. If that decision is around 95% unconscious, you must ask yourself to what extent you're in charge of your life? To what extent is your subconscious mind, running mostly on automatic, determining what you focus on and therefore what you manifest?

The question is whether you can integrate the conscious and unconscious functions of your mind to generate a more effective cognitive team to accomplish your goals. When left on its own and isolated from conscious awareness, the subconscious mind (as Freud so aptly proved) will often become neurotic and mess up your life. Conversely, when integrated into conscious functioning and given proper direction, the subconscious mind will work remarkably hard to support your conscious intentions.

How do you generate this integration of conscious and unconscious functioning? As therapists have been proving for many decades now, the main tool for communicating your conscious intentions to the unconscious mind is through precise one-liner statements of intent. In earlier chapters, you learned the basic process for communicating with your subconscious mind through positive one-liners. Now it's time to go a bit deeper.

First, you will need to clarify in your conscious mind what you want to ask your subconscious mind to help you manifest. Remember that the subconscious mind operates mostly in computer mode; it is programmed, literally from the womb and birth onward, through experience, learning, and repetition. The power of the conscious mind is it can choose what to focus on, and then communicate that choice to your subconscious mind to further that action.

This is done by talking consciously to yourself. Your subconscious mind is always listening if you ask it to pay attention. It seems to learn best with the insertion of a single sentence, although more lengthy ruminations also seem to impact it. The unconscious function of your mind manages your massive memory storage, your vast associative functions, your rich symbolic images, and your full range of emotional responses. In contrast with your conscious rational mind, the subconscious mind isn't a great thinker. It needs to be spoken to simply, positively, and with clear intent.

The subconscious contains so many things that have nothing to do with thinking logically. The subconscious takes things quite literally and constantly works at making everything you say and believe true for you, especially the things you say about yourself. What we believe is all-powerful in the subconscious mind, and to make our inner thoughts true, the subconscious mind must do some rearranging. It does not have to make logical sense to the subconscious; it only has to create a connection in the mind so the belief can hold to be accepted. Usually this is done though repetition or when something dramatic happens and you make an unwavering decision with every cell of your body, mind, and soul. Most of the time, the connection is accepted because the subconscious is trying to avoid pain or trying to gain pleasure. When training the subconscious mind to do what we want, we need to know its rules and how they work for us.

One of the best things about life is that we get to make our own rules; unfortunately, we unconsciously adopt so many of them because somebody told us those are the rules we need to believe. Remember the monkeys and the banana story?

Right now, what would you like to say to your unconscious mind, to let it know what you want it to help you with? This could be reversing any obsolete rules or creating a new belief or rule that serves your new mission, dreams, and goals. Let's pause a moment so you can explore this process. Tune into your breaths coming and going, so that you relax and quiet your mind ... see how you feel when you say to your unconscious mind:

"I want us to work together as a great team."

Now you can experiment further by saying to your subconscious realms:

"I respect you and need your help, please."

Perhaps you might also see how you feel when you say to your subconscious presence:

"Please give me the power and insight to achieve my goals."

THE MYLAR PATHWAY

We've been talking here mostly about the mind, which is somewhat different than talking about the brain. The mind isn't really a scientific term; traditional neuroscientists always talk about the brain, not the mind. What's the difference? Your brain is the material organ in your head, part of the tangible world of your physical body. Your mind, in contrast, is part of the invisible transcendent realm of your thoughts, your emotions, your attitudes, and your beliefs and imaginations.

Your brain is the obvious physical organ that's usually associated with mind and consciousness, but the mind isn't necessarily confined to the brain. In fact, the intelligence of your mind seems to permeate every cell of your body, not just your brain cells. When we talk about your mind, we're talking about quantum realms where matter in the Newtonian perspective is transformed into energetic relationships that transcend Newtonian laws.

The subconscious-versus-conscious discussion we just moved through is a phenomenon of the mind, not just the brain. Neuroscientists can observe areas of the brain that light up in brain scans when the conscious mind seems to be activated, but when we talk about consciousness itself, we're entering an aspect of reality that is not confined to the physical brain. In fact, we've seen that in quantum theory the whole universe can be seen as a conscious mind that we're somehow plugged into.

Now I want to share insights drawn from physical brain science that shed light on how you can actively integrate the unconscious and conscious dimensions. Specifically, what's happening in the brain when you repeat a positive one-liner over and over to yourself? Why is this so effective?

When you say something over and over again, it's now proven that the brain responds to this action by developing new and very specific neural pathways that somehow encode and hold the thought you're thinking as a new neural expression in your brain. You literally impress a thought physically into your brain through repetition!

Recent neurological research clearly documents that with every new activity or novel idea you focus on, you initiate a new neural region of growth in your brain. This has been happening as you read this book. You're growing new neuron paths as you entertain a new idea. Each time we return to focus on that idea, you'll stimulate more neural growth in that region, and extend neural connections into other brain regions, creating links and associations that didn't exist before.

In terms of mindset expansion, you can actively acquire new physical real estate in your brain that's devoted to the mindset you're nurturing. This is clearly an act of purposeful creation! Your almighty power of attention, combined with the conscious act of thinking and speaking particular words and ideas, transforms regions of your brain in the directions you choose.

As you continue focusing on particular thoughts, ideas, and actions, as we've been encouraging, each related nerve-cell (axon) in your brain will develop a myelin coating which enables the flow of electrical energy and coded information to occur faster in that region, with much less energy expenditure.

Also, with regular repetition of activity, these well-developed and "paved" neurological systems can become habitual patterns that function almost automatically. Here we can see how something deposited in the brain through repetition can shift from conscious mental activity (that flows relatively slowly) to unconscious activity (that flows much faster). Over time, repetition helps mental habits develop and generates brain regions where new beliefs evolve and become stronger.

This process of progressive myelin coating helps the brain function more efficiently, but it also makes it harder to let go of old habits and beliefs. Curiously, when you regularly meditate and quiet the mind, it's now documented that you increase many positive brain conditions, including the acceleration of myelin coating of the neural cells. Simple mindfulness meditation mysteriously generates much-improved inner balance, harmony, and integration between conscious and unconscious functions of the mind.

Neuroplasticity is a related brain feature. Also referred to as brain plasticity, neuroplasticity is the ability of the brain to alter and adjust its functioning over time. Because of the brain's neuroplastic capabilities, brain activity associated with a particular function can be transferred to another location; the proportion of grey matter can change; and synapses can strengthen or weaken over time. In other words, the brain isn't hard-wired; it's flexible, it changes over time, and even as an adult you can transform your brain by shifting your focus, habits, and taking in new information and experiences.

When you perform activities or repeat one-liners, that region of your brain will change and expand. Conversely, when you stop doing something, your synaptic potential in that region will decline, and material from that region will be used elsewhere. Right now, you have the freedom and power to make your brain work in new ways for your survival and heightened fulfillment.

This is a great liberating fact. The one-liners imposed on you early in childhood aren't set in stone. When you expand your understanding of how the universe works, you activate your neuroplastic capacity to let go of the old model and embrace a radical new mindset. In essence, you can self-evolve and literally, inside your own brain, become a new being.

Bruce Lee, when talking about his famous kick, said this about establishing a new habit and becoming great: "At first a kick was just a kick, but then I broke down the components of a kick and realized all the little nuances, and it became very complicated. Then, after I practiced the kick enough, over and over, I became truly great ... and my kick just became a kick again."

"I fear not the man who has practiced 10,000 kicks once, but I fear the man who has practiced one kick 10,000 times."

—Bruce Lee

What he's saying is that he practiced his kick so many times he didn't have to think about it anymore. He just thought "kick" and he got what he wanted. The desired action moved from being conscious to mostly unconscious and became much faster and automatic. That's how it is in our everyday life. The body and mind want to use the path of least resistance and least energy. When you do something repeatedly over time, it downloads to the subconscious and becomes second nature, making it automatic with little to no effort. It *seems* effortless because that's the goal of the body and mind when working in unison.

We work at our chosen activities and practice them over and over until a new circuitry is formed, until the brainwave connection has become so engrained and instantaneous that we barely think of it. It immediately happens.

First of all, we must do the work. Our conscious mind is a great tool, but it doesn't run our lives. Our conscious mind sets the goal, and the subconscious mind hits the target. We must consciously determine what we want and need so we can make our life the greatest it can be. You can set your goals big or small; just remember, it's the success of achieving each goal that builds confidence and momentum to achieve the next one more easily.

SET THE BAR LOW AND THE TARGET CLOSE

Tony Robbins was hired to help the military train people to shoot. Tony had never shot a gun before, and you may ask how can someone who had never shot a gun train experts on how to shoot?

He decided he needed to help the sharpshooters eliminate any self-doubt they may have adopted about their marksmanship. He had them bring the targets ridiculously closer than normal so there was no way anyone could miss. After repeated success, he had them move the target back a few feet, still close enough that it was child's play for them. He progressively moved the target back, building up the group's confidence so there was no longer any threat in missing the target. They repeated the exercise until they believed they were going to hit the mark every time with complete and total conviction.

By making the target impossible to miss, Tony gave the sharpshooters the confidence they needed to eradicate their self-doubt. This training improved the graduates so much that they set a new record that year.

As in the training, you may want to start with easy goals to train the subconscious. You can have any goal you want, just make sure your goal is not too far from what you already believe. Once you have goals, you'll need to review what your rules are and make sure both your rules and goals are in alignment.

Review your beliefs about yourself and the rules you made about life over the years. Then decide what you would need to believe for you to achieve the goals you are setting then create new rules. Find out what doesn't align and do subconscious training to change it.

To be good at something, we must break down its components. We must exercise the pattern over and over until we master it. Only then can we relax, shift "into the zone," and experience our life flowing spontaneously. Being "in the zone" is a quality of consciousness that reflects the quantum mindset. We've all seen quarterbacks who are so tuned into their wide receiver they can throw a football fifty yards into the hands of a player who's running full speed. Once a quarterback perfects throwing a football, by game-time he relies on his subconscious instincts to hit the target. When everything is moving 100 MPH and he's being blitzed by 250-pound muscle men rushing at him, there's very little to no time to think about what goes into throwing a perfect pass. They must practice it so many times that it downloads into their subconscious mind and becomes a natural instinct.

This is the same process you need to do to be successful and achieve your goals. You must train your subconscious to automatically think the positive things you want. You must constantly be aware of the negative one-liners you say to yourself and immediately reverse and replace them with positive one-liners until they become your default mindset. If you train your subconscious to totally believe in the things you want, you will hit the target you're shooting at a lot more times and attract to you what you're wanting to manifest. Even when you think the road is taking you down a different path than you intended, have faith that you will still end up at your destination, or a place even better.

EXPRESSING YOUR TRUE INTENT

Before mastering a desired intent and manifesting your chosen outcome, you're going to need to get quite clear on what you want to accomplish. You'll also need to deal with any subconscious resistance to accomplishing your goal. First, establish and clearly state your goal, then transcend negative one-liners by flooding your brain with an equal and opposite statement of positive intent.

There's a strong and valid therapy movement now in vogue called self-talk therapy. It's become so popular recently because, well—it works.

We all talk to ourselves, and much of this self-talk runs through the back of our minds at mostly unconscious levels. Our subconscious seems to almost constantly remind us of who we are, what we want, what we're afraid of, and how we want the world to see us.

If you listen to this ongoing internal monologue, you'll find that you're thinking either negative or positive one-liners about who you are, how you feel about yourself, how you feel about your ability to achieve your goals, how you want to be viewed by the world, and how you relate to people. Your internal monologue determines the quality of your self-esteem, your self-confidence, and your self-image.

The key question is: Are you immersing yourself in negative or positive one-liners? Are you knocking yourself down with your self-talk, or boosting yourself up? Don't be surprised if a lot of your self-talk is negative, and don't be surprised if that negative inner voice sounds a lot like your early-childhood caregivers.

Parents unwittingly pass onto their children their own inherited negative one-liners; that's just how society works. You took into your subconscious mind what Freud called the super-ego, that part of your mind that acts as a self-critical conscience, reflecting social standards learned from your parents and teachers. The rules, judgments, attitudes, and condemnations of a parent are almost always ingrained in the subconscious mind of the offspring, usually before the child is old enough (around five or six) to consciously reflect on that programming.

If your parents and teachers were positive and mature, you probably don't have much of a problem with negative self-talk,—but you're in a small minority of lucky people. Most of us go through our entire lives never dealing with that nagging, negative voice that keeps telling us we're no good, we're stupid, we're weak, or we're bad.

The first step in constructive self-talk is to begin observing that inner-critic voice so you bring the mostly-unconscious function of the mind into clear light. As you expose your self-sabotage mechanism, you'll be motivated to do something about it. By using select one-liners and saying them to yourself frequently, you can begin to re-educate your subconscious mind with a new set of defining attitudes, intentions, assumptions, and beliefs. This will change who you are for the better!

If you find your critical inner voice saying, "I'm a bad person," set yourself free from this programming by fighting fire with fire. Start saying consciously to yourself (which includes your subconscious self!), "I'm a good person." Soon you'll find that you begin to go out of your way to prove you're a good person. If you say, "I'm a smart person," you'll do things to prove you're a smart person.

In other words, you're taking charge of your own mind. You consciously focus your full mind's attention on the type of person you want to be, then you begin to act in that positive mode, and everything begins to shift in the direction you want.

What you're doing here is choosing who you want to be, how you want to manifest your very own self into the world, and then you're acting to generate that change. This is the deepest level of manifestation you can achieve. When you transform your self-image at deep levels, your whole life will begin flowing in a positive direction. In short, what you're focusing on right now is determining what you will become. You imagine how you want to be, you declare this intention, and then you act on it.

You change your life's entire trajectory into the future by changing how you describe yourself. By fervently believing in the positive qualities of your higher self, you awaken this belief deep in your subconscious. Then it goes to work to accomplish the stated goal.

Choose to stop calling yourself stupid or lazy or bad or any of those negative things, then act to replace those critical put-downs with positive, self-edifying, one-liner beliefs and updated rules.

That's the process: consciously focus your whole being on what you realistically want to become, repeat this over and over, then enjoy how you step-by-step become who you want to be. This isn't some drifty-dreamy fantasy or New Age illusion; this is the actual scientific process being revealed gradually by our emerging quantum revolution in psychology.

WHAT DO YOU REALLY WANT?

Your thoughts have power. They impact your own inner experience and identity and also the world around you, but only if you are clear about your core needs, and fully believe you have that power to fulfill those needs. Then, by entertaining new thoughts and envisioning the new world you want to live in, you'll gain new power to manifest the world you desire.

I encourage you to begin thinking and acting now as the person you want to be. I remember my kids asking me how to find a special boyfriend or girlfriend, and I told them, "You can't just go out and find one. You need to first think about the type of person you want to be close to with, imagine all the characteristics of that desired friend who makes you feel happy and fulfilled. Then you'll need to think about this: 'Who will I need to be to attract the person I want?' Next, you'll need to start being that person. When you're consistently being the person who would attract your desired friend to you, you can trust they will show up."

Manifestation isn't just about manipulating the outside world to get what you want, even though we're traditionally taught that strategy. Manipulating is a Newtonian approach to life, learning the laws that govern material reality and applying those laws to build things and influence people to your benefit. In that old model, you're the subject, the experimenter, the builder, the boss, and you purposefully manipulate objects and people to get what you want.

At more subtle levels, quantum science and theory has upset this manipulative mindset, because the predictable existence of cause and effect has been challenged. Instead of holding the experimenter outside the experiment, in quantum research the experimenter is a part of the experiment. Rather than *manipulating* objects or people, we now choose to *participate*.

Here's a key point also emphasized in quantum research: We can't manipulate and participate at the same time. We must function from within a mindset using one or the other as its base operational mode. If we're wise, we'll choose to participate, because participating in the greater whole is the deeper nature of how the quantum universe itself works.

A large part of shifting into a quantum mindset involves letting go of seeing yourself as a manipulator; instead, see yourself as a participant in everything you do. The quantum vision is that at subatomic energetic levels, your thoughts do impact the world.

If you say this primal thought over and over to yourself for a few days, "I choose to participate rather than manipulate," your whole being will be transformed in this participatory direction.

This is what happens when you align yourself consciously with the truths we've been exploring about quantum theory. Your dominant thoughts establish the intent and direction you're choosing in your life. Either you let old ingrained one-liners run your life, or you expand your mindset by choosing positive one-liners that express your true intent. Regularly hold these new one-liners in your mind.

Through this self-talk you'll communicate with your potent subconscious mind what you want to accomplish, who you want to be, and what and who you want to attract into your life.

Let's practice this process again. Give yourself a bit of free time right now to take a breather ... settle into your chair, perhaps stretch and yawn a bit ... first focus your attention on your feet ... expand your awareness to include your legs, knees, and thighs ... next, expand your awareness to include your pelvis ... your belly ... your chest ... your neck and mouth ... and your head ... be aware of your whole body at once, here in this present moment ...

Now focus toward your inner center ... in your heart. Breathe into whatever feelings you find there ... begin to ask yourself, "What do I really want to bring into my life?" Say this a number of times on each new exhale ... see what comes to mind ...

TAKING THE LEAP

One of the great breakthroughs in quantum thinking was the proven fact that electrons and waves function energetically in ways that defy old-time Newtonian logic. As Niels Bohr first postulated in 1915, when a single electron becomes entangled with a solitary proton, a hydrogen atom is created, where the electron is locked into a particular orbit around the nucleus. If the electron becomes excited energetically, it will release a photon and leap suddenly into a higher orbit. Here's the key point: the electron doesn't move through time and space into the new orbit, it instantly appears in that new orbit, seeming to violate Einstein's law that nothing can travel faster than the speed of light.

Physicists continue to argue over what actually happens in this regard at subatomic levels. Metaphorically, this seemingly-instantaneous leap from one state or position to another reflects how a person shifts into an expanded mindset. First, information comes to a person's current mindset that stimulates an energetic charge. This charge grows to a point when the old order (mindset) can't contain it. Suddenly, a leap happens that isn't a progressive development, but an instantaneous shift into a higher-order state.

From my understanding of psychology, this is exactly how a quantum leap into an expanded mindset happens. You input new and sometimes startling information such as I'm providing in this book. You build up an energetic charge that begins pushing at the walls of your current mindset for a while, then suddenly a leap of consciousness occurs.

Leaping is the letting go of where you are, putting away the known, and committing yourself to embracing whatever awaits you in the new realm. You choose to move into a new energetic mental state, and to feel a new sense of engagement with your core being. When you take this leap, you *will* feel transformed—you've gained a new perspective on life, seemingly all at once.

In this new perspective, you will have the mental power to purposefully reprogram your mind to more effectively generate the life you want. You become a mind pioneer well-equipped with the mental capacity to make quantum leaps into ever-new, more-enlightened states of mind where the impossible becomes suddenly possible.

Some say there's a place in the Universe where everything and nothing exists at the same time. It's a vast, dark place in outer space of pure nothingness, yet it's the place where all possibilities and probabilities live in the Universe. The vibrational field is known as The Void, or Point Zero Field, which is the base where everything sits in the mathematical field of quantum physics.

Meditation is the practice of stillness and coming to a place of equanimity. Maybe this place of deep stillness and nothingness is why mediation is so powerful and why sages and gurus advocate for it. Maybe meditation is what enables us to tap into the consciousness of the Point Zero Field, where all possibilities live and where we can access our full manifestation powers.

Understanding life through new quantum insights will set you free to live in a vastly different realm than you originally thought possible. When you entertain new thoughts and envision the world you want to live in, you naturally gain new power to manifest the world you desire. In this spirit, I encourage you to begin thinking and acting as the person you truly want to be, right now—today!

Imagine you're entering a room full of people you don't know—you don't even know why they're together. As you walk through the door and people turn to look at you, who do you want them to see? If you set yourself free to be the real you, who is it you want to be?

As you walk down the center aisle of the room with people watching you from both sides, notice how you walk when you set yourself free to be the real you. What's your gait? What's your posture? What's the overall impression you're broadcasting outward?

As you sit in an open seat, how do you want to feel? What's the core dominant feeling in your heart that makes you ... you? Who do you want to be, as a feeling that people can tune into and receive as your deeper individual essence?

Chapter 9:

MANIFESTING YOUR DREAM

We now have all the ingredients we need to lay out a realistic quantum manifestation program. I want to offer you a specific process for the primary manifestation goals most people delve into at some point in their lives. This chapter will deal with your personal needs and aspirations, and the last chapter will explore how to use your personal manifestation powers to help the world in general.

Let's quickly review several key mindset points before moving into action mode. Throughout this book, you've been gradually developing an expanded quantum perspective that will enable you to see more clearly how your inner emotions, thoughts, and imaginations are constantly interacting with and changing the world around you. By learning to observe your subconscious mind at work, you're in the process of integrating your subterranean functions with conscious intent and future vision.

As we've seen, when aligned with conscious goals and positive emotions, your subconscious mind is the source of your true power, because its foundational beliefs create and channel your body's core emotive energy that manifests and manages much of your personal world. It's to your advantage to continue in the months to come to get to know your subconscious function in more depth, clarity, and appreciation.

When your subconscious mind becomes aligned with your conscious intent and the laws of quantum science, you gain more power to fully express who you really are in the world. You'll find you can respond instantaneously with integrity and authenticity to every new moment, as you act harmoniously to help manifest a world where inner and outer realms are united.

Always remember your subconscious mind runs 40,000 times faster than your conscious mind yet uses less energy. It is an important rule the body and mind have because one of its main objectives is to conserve energy. While you're thinking one thought consciously for half a minute, your subconscious mind is running millions and even billions of tiny bodily operations, all at the same time. It's also continually making subconscious decisions and acting on those decisions at cellular and subatomic levels.

If you think something cannot be done, your subconscious will tell you, "It can't be done, so why waste the energy?" This misalignment is the root cause of procrastination and self-sabotage. If you think you can, but you're afraid, you might try, but the conflicting energy of worrying is getting in your own way. With no real focus, the mind and body will determine for itself the energy it needs to accomplish a task or goal. If you haven't practiced creating a solid belief that you can do something, when any real effort is needed to accomplish that task, you'll automatically sabotage yourself. If you totally believe you can do it with all your heart, then it will seem effortless.

> *"Whether you think you can, or you think you can't—you're right."*
> —Henry Ford

Change takes energy; the mind and body will fight it, but with persistence and the right thinking, circumstances can and will change. You must believe with all your heart and celebrate as if you already have achieved what you want; this is when the subconscious will take over and just do it.

In the Bible, it says, "Whatever you ask for in prayer believe that you have received it, and it will be yours." (Mark 11:24). You must believe as if it has already been given to you or has happened already. Many religions have similar statements. They say if you believe fully with all your heart and soul, those things will come to you, one way or another.

The ego thinks it's in control, but in reality, the vast bio-robot functioning of the subconscious mind is deciding almost everything you're doing, moment to moment. Therefore, it's wise to put that subconscious power and organic wisdom to work on what's most important to you.

Consider yourself in the process of entering into this higher level of integrated awareness. As you gradually integrate the conscious and unconscious functions of your mind, you'll naturally advance into a more powerful and trusting mindset. You'll find that your enhanced observational and manifestation powers enable you to participate more fully within the unified quantum reality.

Most importantly for manifestation, we've seen that non-physical consciousness seems to function outside the space-time continuum; Universal consciousness is the non-material energetic matrix within which the space-time continuum operates. Perhaps the greatest leap into a quantum mindset happens when you realize things can happen simultaneously and spontaneously at non-material levels.

Logically, this means that your thoughts and your emotions can radiate beyond your local sensory presence and influence the world around you at quantum levels, just as the Princeton PEAR studies documented. You can broadcast your intent at quantum levels, then act physically to attain your goal. That's the quantum vision, so let's take the leap and begin practicing the basic quantum manifestation process, applied to a number of different goals and intentions.

START WITH THE EASY STUFF

People often start on the wrong foot and try to manifest the biggest thing they need or desire in their lives right away. My advice is to start by consciously manifesting very simple things that you believe and know you can readily accomplish, rather than big complicated ones. First exercise your manifestation muscles by fulfilling minor intentions. Walk your subconscious mind through easily-completed expressions of the same process you'll then apply later to bigger goals. Establish a solid habit of this process through minor successes, so your subconscious mind can firmly understand how you're wanting to work with it. As this process manifests, you will simultaneously gain confidence as you strengthen your belief that this process works and you can use it for anything.

In short form, the primary manifestation model we're using is this:

1: Look inward and observe your feelings, your ideas, and your motivations regarding what you want to accomplish.

2: Decide and declare your intention to accomplish that goal.

3: Imagine succeeding with that goal and reinforce your belief that you can accomplish the intention by feeling as if it's already accomplished.

4: Act on your inner vision and intention, merging imagination and vision with the physical reality around you.

5: Here's a key step! After you've set things in motion, choose to participate rather than manipulate the outcome of the process you've initiated. Surrender to the outcome that naturally emerges, not just through your ego-will, but through the higher will of the matrix in which you're participating.

6: Finally, once you've completed the first five steps, take time to reflect on the process, and express thankfulness for how the universe has participated with you to accomplish your desired goal.

Here's the process in a tighter format:

1: **Observe** *your current needs and yearnings.*

2: **Declare** *your intention to manifest a particular goal.*

3: **Believe** *in and imagine accomplishing that goal.*

4: **Act** *to physically manifest your goal.*

5: **Accept** *the natural outcome.*

6: **Reflect** *and express thankfulness.*

Observe ... Declare ... Believe ... Act ... Accept ... Reflect

You can apply the same six-step process to manifesting an ice cream cone or a life partner, a new coat or a new job, a novel invention or a great vacation; the core process is the same. I'll also provide you with an audio guidance program on my website www.TheQuantumMindset.com, so whenever you feel the need or desire to manifest something in your life, you can relax and let the audio program guide you through this process. After a few times, you'll memorize the six steps and move through this process on your own.

Exercise: Manifesting Made Simple

Let's move through this together by manifesting something simple. Several times every day for several weeks, I recommend you exercise your quantum mind by applying this success formula to little things you're going to do anyway and that you know you can accomplish. Do this over and over so your subconscious mind develops a reflex habit that kicks into gear whenever you face a more challenging manifestation goal.

Imagine that it's hot and you want an ice cream cone (or pick another simple goal of your choice). In the next days and weeks, run variations on this simple-need theme:

Observe ... *pay attention to the initial hunger or need or yearning or desire you feel—experience the bodily motivation of wanting that ice cream cone! Notice where the desire for the ice cream cone originates. Your body is hot, and you want to cool off. You remember eating an ice cream cone in the past, and how it satisfied your need ... you'll feel better once you attain this simple goal ...*

Declare ... *after determining the inner source of your desire or need, make up your mind and declare to yourself your intention to manifest that ice cream cone. Say clearly to your subconscious mind, "I want to get and eat an ice cream cone."*

Believe ... take time to imagine exactly what you're going to do to get your ice cream cone. Use memories and imaginations to envision you acting to accomplish your goal—say to your subconscious mind, "I believe I can do this!" Feel the confidence that arises when you reinforce your self-esteem.

Act ... go into physical action and do it! Perhaps you get up and walk to your refrigerator, open the freezer section, and take out an ice cream cone.

Accept ... as you interact with the world around you to manifest your intent, be sure to participate rather than manipulate the situation, by accepting the reality you discover, and surrendering to that reality rather than fighting against it. After all, no one fights against reality and wins! If you open your freezer and discover that someone already ate your ice cream cone, willingly surrender to that fact, or go out for the ice cream.

Reflect ... after you accomplish the goal, hopefully getting and eating your ice cream cone, remember all five steps you just took to accomplish your goal.

... Impress into your subconscious mind that you want this process to become a new habit ... feel thankful to the wholeness of the universe for participating with you in satisfying your need or desire.

FOUR CORE MANIFESTATION GOALS

Each of us is unique in the details of our lives. At the same time, most human beings share the same set of primary needs, desires, hungers, and intentions. While there are variations, listed below are several universal human needs, beyond the basic physiological requirements:

Safety / Security / Money

Love / Belonging / Relationship

Self-Worth / Esteem / Ego Strength

Fulfillment / Realization / Wholeness

Let's look at each of the four primary types of human need and intention, to see how they relate to our previous quantum-mindset discussions. Then we'll round out this chapter by applying our Six-step Quantum Manifestation Process with examples of each.

Safety / Security / Money

We live in a physical world full of material things we want to have for our own. Many people, when asked what they want to manifest in their lives, focus on physical things such as a new car, computer, or cell phone, or perhaps a new house in a safe neighborhood, a hefty pile of cash in the bank or stock in hand, or any other physical thing they think will make their lives better.

There's nothing wrong with focusing on this physical level of manifestation, as long as you aren't running on a buried one-liner belief saying, "If only I get that thing, I'll be satisfied." Things don't necessarily bring happiness, but they do help. If the basics of safety, security, and money aren't taken care of, life can be quite unsatisfying.

Love / Belonging / Relationship

This second type of need is often placed front and center on a person's need list. Most of us need to feel loved and to share love; we yearn to belong to a family, a community, a country, and perhaps a church, a corporation, or a sports team. One of our few human genetic fears is feeling abandoned, expelled from a family, community, or business.

Our human hearts seem programmed to hunger for close relationship. Most of us grew up in a family, and we are automatically programmed with universal, seriously-deep and wonderful yearnings to feel we belong, that we're loved for who we are, and accepted and included in a family circle and social network that sustains and nurtures us. According to a recent survey, in 2016 there were 35 million people living alone in America, 28% of all households (that's up from 17% in 1970).

Manifesting fulfilling relationships is a primary intention we can address with our Quantum Manifestation approach. There are also specialized books and programs applying similar methods to this particular human need. Non-intimate friendships and business relationships can also be attracted into your life through this same approach.

Self-Worth / Esteem / Ego Strength

Often, succeeding with the first two types of human needs can be chronically thwarted because a more inward need is not yet satisfied. A primary need is feeling that deep-down you're okay, you're adequate, you're of value, and respected for who you really are. Rather than wanting to manifest a thing or a person into your life, the need is to manifest an inner belief and feeling of wholeness, personal power, ego health, and related self-esteem factors.

We've talked quite a bit about this third, nonmaterial aspect of quantum manifestation, especially in our discussions of how ingrained negative beliefs, attitudes, and one-liners dwelling in our subconscious realms can sabotage our ego strength. You'll want to revisit those sections of this book and regularly work on replacing the critical inner voice with positive one-liners and attitudes.

Fulfillment / Realization / Wholeness

Most people in all cultures also sense a need for developing a world view that has deep meaning, that provides a feeling of being fully satisfied in life. We seem to naturally hunger for an integrated understanding of what life is all about, a belief system that's expansive and rings true deep down in our hearts, minds, and souls. We want to feel one with God, with Great Spirit, with Yahweh, Buddha, or Lao Tzu or in quantum realms, with the conscious, compassionate, all-powerful, and all-inclusive universe of which we're an integral part.

As in the first three types of manifestation, this fourth type of human "yearning for manifestation" is equally applicable to the Quantum Manifestation Process. I've found satisfying comfort and meaning in life by expanding my own mindset into quantum realms, and I hope this book has inspired in your heart and mind a similar sense of enduring wholeness, hope, opportunity, and loving entanglement.

APPLYING THE Q/M PROCESS

As I said earlier, you can apply the Quantum Manifestation (Q/M) Process to any type of intention or desire. The more you move through the process, the better and faster you'll be at doing it. Let's round off this chapter by clarifying what your particular needs are right now, by using positive one-liners that you can then act on at your leisure. Here's the basic needs list again, for you to work from:

Safety / Security / Money

Love / Belonging / Relationship

Self-Worth / Esteem / Ego Strength

Fulfillment / Realization / Wholeness

Let's consider what general focus phrase works best for each of these four categories, so your subconscious mind knows your intent. How about these clear positive one-liners for the first category:

Safety / Security / Money

"I need to manifest a situation for my life where I feel secure and prosperous."

Is this a need of yours? Or perhaps:

"I need to manifest enough steady income so I don't worry about money."

Does this ring true? If so, apply the basic Q/M Process to fine-tune what you want to manifest, then imagine and act on a game plan to fulfill your need. If you spend enough time and loving attention on the first few steps of the process, your actions will tend to flow naturally to fruition.

Regularly return to the quantum insights of this book to sustain your expanded world view and tap into the higher integrative powers we've explored. Also, examine your one-liner attitudes about money to identify the negative ones that hold you back and replace them with positive attitudes that will motivate you to act to fulfill your needs. This is a vital focus and sometimes takes considerable time to move through, so be patient with your flow.

Love / Belonging / Relationship

*"I want to become close with someone
special who fulfills my need for love."*

Or …

*"I want to join a group of people where
I feel I belong and am accepted."*

Or …

*"I want to attract into my life
someone I can work with harmoniously."*

These are just a few general examples to stimulate your mind to come up with specific statements of intent to move through the Q/M Process successfully. Feel free to play with them however you want. Perhaps get out a pen and paper or tablet or whatever you use to make notes and set your creative process free to explore what words most clearly express your present need in this category. Let your subconscious mind participate in the process. Remember to pause and tune into your breathing and your whole-body presence, as that always opens doors into deeper reflection. Simply ask yourself, "What do I really need?"

Self-Worth / Esteem / Ego Strength

*"I want to feel I'm valuable, respected,
and needed by my friends and colleagues."*

Or …

*"I want to strengthen my sense of confidence
and ability to accomplish great things."*

Or ...

*"I choose to honor and love myself
and receive praise and support."*

Again, you'll want to turn to your subconscious for help in fulfilling these needs. Negative one-liners lie at the heart of most people's problems with self-esteem and ego strength. You might find that at the heart of your sense of low self-worth is the gnarly one-liners, "I'm no good," "Nobody likes me," or, "I'll never get ahead."

Consider various negative one-liners until you find one or more that are especially strong in your subconscious mind, then apply the methods you've learned in this book to replace them with positive one-liners such as, "I'm a good person." or, "I'm a good friend." or, "I'm confident I'm going to succeed."

Say your positive one-liners often to yourself, write them down, and perhaps record them and listen to them. Do all you can to reprogram your subconscious mind with your new chosen attitudes. Remember, if you need some professional help in this regard, it's a sign of strength, not weakness, to reach out for help when you need it! Cognitive therapy, first created by Arnold Beck in the 1960s, has come a long way and is effective, rapid, and uplifting.

You'll go a long way on your own with focus phrases such as, "I honor and love myself, just as I am." When you say such a focus phrase to yourself, many times throughout the day, you reprogram your subconscious mind in exactly the direction you want. I highly recommend this process, but it's up to you to do it!

Try coming up with a simple self-esteem one-liner first, such as, "All day today I'm going to remind myself that I'm perfectly okay and loveable." Set this as an inner action you're going to manifest.

At least half of what you'll want to manifest in your life won't be physical, a thing, person, or situation—it will be a new inner quality of mind. You might say to yourself, "All day today I'm going to see the world in a new, more-positive light." Move through the six-step Q/M Process with this intention and see how you do.

Fulfillment / Realization / Wholeness

"I choose to feel satisfied and fulfilled in my present life."

Or ...

"I open my heart and mind to receive new insights and realizations."

Or ...

"I want to experience feeling whole and one with the universe."

This fourth dimension of manifestation brings us to the theme of meditation, reflection, contemplation, and spiritual awakening. Do you feel any yearning in this direction? Some people do, and some people don't. Even if you're entirely non-religious, I hope you've seen from this book's quantum discussions that just being alive and conscious right here right now, in a universe that's seemingly infinite and very possibly a unified conscious presence in and of itself, is a deep spiritual experience.

You don't need any formal meditation program, religious training, or spiritual group to fulfill yourself at this level of "meaning and awakening." All you need is what we've been exploring in this book, an open mind and heart that's willing to release outmoded beliefs and limitations and leap into newness of mind and spirit.

It does seem like the universe welcomes everyone to take this leap! We're all being encouraged to let our hearts, minds, and souls become entangled in the greater web or matrix of the universe. Your challenge is to develop a mindset that embraces rather than judges or rejects the higher realms that quantum physics reveals.

The good news is, taking the quantum leap seems to be a fully natural, safe, and rewarding thing to do. All you do is set your intention to take the leap, then follow the Q/M Process to accomplish this goal. You'll find additional support programs on our website to assist you in this intention. (www.TheQuantumMindset.com)

To end this chapter, perhaps you'd like to ease up, put the book aside after reading these words, and see what pops into your mind regarding what you want to manifest in your life. Let yourself take a breather ... tune into the feelings throughout your body ... the energy flows ... your sense of earthly balance ... feel the invisible air flowing in and out of your nose or mouth ... and the whole-body sense of "being fully here right now" ...

Expand your awareness to include whatever feelings you find in your heart ... breathe into these feelings and let them flow ... open your awareness to whatever's rising up into your mind from your subconscious realms ... say to yourself:

"I'm open to listen to the needs
and wisdom of my heart."

As you stay tuned into your thoughts and feelings, see what words come to mind when you ask your deeper self:

"What do I really need to manifest
in order to feel content and fulfilled?"

Chapter 10:

CHANGING THE WORLD

We've explored quantum-level ways to help you develop an expanded mindset that will empower your life. We've also seen how you can apply your quantum mindset to focus on and manifest what you need and desire in your personal world. In this final chapter, we're going to consider what you can do at quantum levels to help the world wake up and come together to manifest a better world community.

Quantum theory insists we start taking our nonphysical thoughts, imaginations, and intentions as seriously as we take our physical actions. They do impact the world around us at subatomic levels of energetic communication. Rather than this being an esoteric religious notion, it's a clear humanitarian responsibility. Every moment of our lives, we're all continually broadcasting either negative or positive thoughts and feelings. It's up to us!

As we discussed earlier, the PEAR studies at Princeton University's Engineering Department, run by Professor Jahn, statistically documented over a span of three decades how the power of silent focused thought and intent can impact the functioning of a random numbers generator. Subjects could also communicate simple shape and numbers to a target mind at a distance. The impact recorded on sensitive measuring devices was tiny, but of course quantum effects by their subatomic nature are quite small.

We have the beginning of research to document the "mind over matter" controversy that's been raging for centuries among philosophers and scientists worldwide. We should also consider the resonant power of our emotional presence and intent. Can one person or a group consciously broadcast positive emotions and intentions out to the world, and have an impact? People following the spiritual teacher Maharishi did a number of experiments to see if this broader social power of focused intent could be measured.

One of these experiments used several thousand of Maharishi's followers to demonstrate the quantum power of "unified focusing on good." They organized 7,000 meditators to focus daily on encouraging the world to become a better place, more peaceful, cooperative, and compassionate. They did this simple meditation for five minutes a day. At the end of the experiment, they evaluated statistically how various measurable situations in the world had changed.

Overall, they found significant changes had occurred. The total number of deaths worldwide went down, all types of terrorist acts were reduced for that period, and newspaper reports of violence were fewer. The complex study indicated the world became substantially more peaceful during that specified meditation period. That was only 7,000 people focusing and thinking together about good, positive things for the world. A similar study in Washington, D.C. with 4,000 meditators produced statistics of a 15% reduction in reported violent crime during the period of the "pray for peace" group meditation study.

At quantum levels we see how the whole universe runs on coded energetic waves that permeate all matter. We've learned to respect the fact that our thoughts and feelings are immaterial but equally "real" compared with material realms of reality. I'm not asking you to believe in the Maharishi's studies, although they seem quite substantial. I'm just encouraging you to see how you feel in your heart about your own power to broadcast good vibes and have them radiate outward and help the world.

Imagine if we could align and coordinate people in much larger numbers to spend just five minutes a day thinking positive thoughts about the world, actively broadcasting at the quantum level a hopeful, healing intent. What could a million rather than seven thousand people achieve? What might a *billion* minds generate?

With proper planning, we could find 25% of the world's population who would want to do a shared "peace" broadcast, to nudge the human population toward more positive feelings and actions. The issue is unity. A number of times already, groups have acted in this positive direction with worldwide prayer initiatives. For instance, a campus Christian organization sponsored a "prayer for peace" initiative just before Gorbachev agreed in 1987 to move Russia toward Glasnost.

These mass social experiments are in their infancy, and hopefully, through books such as this one, more and more people will consider using their individual mental and emotional power to raise the "peace and compassion" vibes of the world at large. Of course, you can do this yourself on your own each day, spending just a few minutes quietly broadcasting a positive one-liner into the world.

It's important for more people to achieve the quantum mindset where they truly believe and know in their minds and hearts that their thoughts do impact matter; prayers for peace, fairness, and prosperity for all have genuine power. Imagine what could happen! On a large organized scale, we could advance our world society beyond ingrained negative thoughts and feelings into an era of lovingly-entangled hearts and cooperative actions.

Only when we believe something is possible can we muster the power and intent required to perform that intent to make a difference in the world. So, keep an eye on how your personal beliefs are evolving in this regard, and see if you can consciously evolve in the direction of believing goodness can in fact prevail.

"Be the change that you wish to see in the world."
—Mahatma Gandhi

40 YEARS OF ZEN—FORGIVENESS RESEARCH

For a number of years, university research teams have been taking a close look at the brains of people who meditate regularly, eliciting such headlines as, "Can meditation change your mind?" Brain researchers have now proven the answer is yes. During a recent conference in Denver called the International Symposium for Contemplative Studies, experts such as Harvard's Jon Kabat-Zinn discussed EEG and FMRI brain studies of Buddhist monks conducted on several American campuses. The findings reveal that regular mindfulness meditation (such as focusing inward with compassion, acceptance, and hope) alters the functioning and structure of the brain in positive directions.

For instance, researchers in Wisconsin hooked up an experienced Buddhist monk to an EEG biofeedback machine and recorded the monk's brain waves during various mental activities and in deep meditation. The assumption was that an experienced meditator's EEGs would be ideal brainwave samples for quantifying optimum human brain functioning. This is called biohacking or reverse engineering. The larger intent of the research was to document that, by using biofeedback training, everyday people everywhere can learn to shift into these optimum mindsets at will.

In the experiment, researchers found the EEG activity of a meditator is quite different from average American non-meditator EEG activity.

Also, they found a definite correlation between regular meditation on positive emotive themes, and brainwave patterns associated with high levels of compassion and forgiveness. In other words, if you want to change the world by changing your brain, meditation is highly advised.

> *"If you want to awaken all of humanity, then awaken all of yourself. If you want to eliminate the suffering in the world, then eliminate all that is dark and negative in yourself. Truly, the greatest gift you have to give is that of your own self-transformation."*
>
> —Lao Tzu

To shift into a monk-like brainwave pattern, first we need to deal with old emotional charges of negativity that directly block us from entering the quantum mindset of meditation. As noted throughout this book, to change the world we must first evolve our own mindset. To do that, we must regularly take time to let go of habits of chronic judgment.

We must forgive everyone we've ever held a grudge against or has been mean to us, and we must also forgive and accept ourselves. Ancient tribal forgiveness practices such as the Hawaiian Ho'oponopono ceremony can be of great value to us all. The powerful nonviolent movement of Marshall Rosenberg called Compassionate Communication offers great tools you can apply immediately.

At the Hawaii State Hospital, there was a special ward for the criminals who were mentally ill. They had committed extremely serious crimes such as murder, rape, kidnapping, or other violent crimes. Criminals would be booked here for two reasons: either they had a very serious mental disorder, or they needed to be evaluated to see if they were mentally stable enough to stand trial.

According to a nurse who worked there, every day a patient-inmate would attack a member of the staff or another inmate. She said the physical conditions of the ward were horrible and the staff were constantly in fear of being attacked, even though all the patient-inmates were shackled. Nurses, wardens, and employees took frequent leaves of absence and were dropping like flies.

Dr. Stanley Hew Len, a clinical psychologist, came into the picture. The staff thought Dr. Hew Len would be the same as other new recruits and act like the savior trying to implement his new theories and proposals on how he could fix the situation, but he didn't. In fact, he didn't do much at all but walk around cheerfully, smile, and occasionally ask for the files of the inmates.

He never even asked to see the inmates one-on-one. Apparently, all he did was sit in his office and look at their files from time to time. As time passed, slowly but surely the conditions in the prison improved and the inmates started to change for the greater good. No one knew why. Prisoners gradually started to be released as well.

For the members of the staff who wondered how Dr. Hew Len was contributing to this new turn around, he shared the Ho'oponopono practice with them and what he was doing.

He would sit in his office and look at the patients' files with specific intentions. While perusing the files, he would feel something in his heart; suffering, pain, empathy. He started the healing on himself, because he believed he needed to be healed and not them. Dr. Hew Len addressed the Divine within him and ignited the healing process by saying the words, "I love you. I'm sorry. Please forgive me. Thank you." That's how the patients got better. Dr. Hew Len had the perception that we are all reflections of one another, and everything is our responsibility and no one else's. He later stated, "I was simply healing the part of me that created them."

The Ho'oponopono work was so effective that eventually the majority of the inmates were relocated, and the clinic had to close down.

If Ho'oponopono can help achieve the monumental task of healing an entire ward of insane criminals, imagine what it can do for you in your life.

The act of forgiveness can be monitored and conditioned through biofeedback equipment, even at home. We can use biofeedback learning to enter into and experience the mindset of forgiveness associated with particular biowave patterns. The general rule is: if we want our personal presence to impact the world at large, we first need to actively let go of (decondition) habitually-held negative energy patterns in our brain and body, so we can enter into the Zen mindset associated with positive humanistic qualities.

You might pause here for a few moments before reading further, to see how you feel about all this. Do you believe your personal state of mind can impact the world around you? Are you improving humanity's condition by improving your inner vibes? Do you feel any motivation to devote just five to fifteen minutes a day to develop a more compassionate inner focus and feeling? Do you consider it your responsibility as a quantum citizen of the world to consciously broadcast good vibes outward to raise the compassion quotient of the world at large?

Do you believe that the love you feel in your heart for everyone on this planet, for all the animals and plants, and even the rocks and lakes ... do you believe how you experience your own feelings impacts life on earth? The impact might be at quantum levels, but there are over seven billion of us on this planet, and if each of us does our part, what a wonderful world this will be!

I feel moved to repeat those "believe in yourself" lyrics from the musical *South Pacific*:

"You gotta have a dream—
if you don't have a dream,
how you gonna have a dream come true!"

FROM PENNIES TO MILLIONS

Most people would love to have a lot of money: money equals power, it promises fun and satisfaction, it gives us a sense of safety and security, and it enables us to spend money heartfully to help the world around us. But so many of us were programmed to think that money is dirty, the root of all evil, and "absolute power corrupts absolutely." These negative one-liners hold us back from prosperity.

If your heartful impulse is to be of help in the world, but you subconsciously block your access to making money and sharing it with the world, spend some time identifying all the one-liners that put a negative twist to the whole idea of being prosperous. When you feel the need for more money, remember that if you don't love the pennies, then the whole realm of money will think that you don't love it, and neither pennies nor millions will come to you.

Again, it's a subconscious thing. Your subconscious mind doesn't think logically; it only has pre-programmed reactions to things and situations. It isn't smart enough to know that money is a million dollars and it's also five pennies. It just picks up on your attitude toward pennies, sees your attitude is negative or uninterested, and acts accordingly. No pennies are attracted and gained, ... and by extension, no dollars or millions are either.

Your subconscious mind knows no denomination; money is money, whether it's pennies or dollars or millions. If you poo-poo the pennies, your body will think, "Money isn't something we want," and you won't attract it to you. Show your body, your unconscious mind, that you honor every penny you see on the ground. Pick it up, kiss it, and put it in your pocket. This shows your unconscious that you appreciate money.

It also shows the universe, through physical action, what you want to achieve. You're expressing the feeling of positive relationship toward the thing we call money. Pause and celebrate when money comes to you. Even if it's just a dime or a penny, feel your appreciation for its presence in your life and love it as much as a thousand-dollar check. It's all the same, so don't throw pennies out and don't say, "Oh, you're just a penny."

Instead say, "Ah, you're a lovely beautiful penny, I think pennies are some of the coolest looking coins there are!" By expressing this thought, you're communicating with your subconscious mind and letting the universe know what you respect and want to attract into your life. Love the pennies and celebrate all aspects of success, especially the smallest achievements, because these are the greatest opportunities to show the universe you not only appreciate all money and success, but you welcome more of it. This principle and practice can be applied to anything you want to attract. Express to the universe over and over what you want and need. Prime the pump!

SPREAD LOVE, NOT FEAR

Fear is a core problem for all human beings. Most of our fear-programming is unconscious, felt down low in our gut rather than up high in our brain. Psychology is gaining a slight understanding of the actual process we can use to reprogram subconscious fear-based reactions into conscious expressions of compassion and cooperation.

One way or another, for peace and harmony to reign, we need to take charge of our thoughts and feelings so deep in our psyche we're resonating with positive expectations rather than fearful reactions. Specifically, we must transform our gut reactions, so we stop habitually looking around us for trouble and problems and start focusing more on generating positive thoughts and expectations.

We find what we expect to find. We need to learn the power of liberated intent to manage our ongoing focus of attention. For instance, we need to focus steadily on *what we want* in life, rather than what we're afraid might happen. That's how we can best act to transform the world.

Unfortunately, most of us fixate *more* on fearful thoughts rather than less. Parents are more fearful for their children now than they were thirty, fifty, or a hundred years ago. Kids are bombarded from birth or even before with the fearful thoughts and apprehensions broadcast by their caregivers. We're all being programmed by the fear-mongering media to be constantly afraid that something unexpected and very bad might happen to us at any moment.

KILL THE MONSTERS WHILE THEY'RE SMALL

The human subconscious is genetically programmed in the amygdala to watch for danger. All animals share this defensive stance, and it has a necessary role to play. It's what keeps us from stepping in front of a bus. In contrast with other animals, we have fertile imaginations that can conjure up horrible images and fearful thoughts that dominate our minds and hold us in a chronic state of anxiety. Fixating on these imagined fears tends to attract similar negative situations in real life.

Negative thoughts are like viruses; if you ignore them, they will live within you. The more you feed these thoughts, the more they grow. It's so important to *kill the monsters while they're small*. Most monsters are built from the imagination and feed off worst-case scenarios, things that seldom occur. Never feed the monsters. Catch and kill them while they're small so they don't grow. Monsters are harder to face when the imagination allows them to grow into something much bigger than they are, merely a figment of the imagination.

I've moved you through this discussion so we can focus with a wider lens on the one-liner thoughts that your subconscious continually runs through the back of your awareness. Such ingrained one-liners are very mysterious creatures, and not at all understood by neuro-science. To transform your mindset, it's important to realize you're dealing with mysterious powers and laws you can participate in, but not fully understand.

Children are most programmable from birth to seven years old. If we start at a very young age to teach our children to look for the positive in life, we could help their brains develop different pat-terns of synapses so they can focus their attention toward positive outcomes and help generate those outcomes. As adults, when we come home from work, we can reprogram our thoughts to let go of thinking, "What did I do wrong today, so I won't do it wrong tomorrow?" Instead we could think, "What did I do right today so I can do it right again tomorrow?"

If we all decide to keep our thoughts and imaginations focused on hopeful positive outcomes, and insist our media seek and broadcast the best and most hopeful things that happened throughout the world, then our own inner broadcasts combined with those of the media would actively reinforce our desired movement into a better world. More and more psychologists such as Rick Hanson at Berke-ley have shed important light on how to override this fear-based programming and consciously rewire the brain with a positive focus.

Studies show our brains are wired to focus 20% of the time on pos-itive thoughts and 80% on negative, fear-based ones. This is serious, and we need to deal with it. We need to set a universal goal to reverse that statistic and look 80% of the time for uplifting content and only 20% for danger.

Could it be a coincidence that most news sources report 80% nega-tive content and only 20% positive, uplifting content? Many people are addicted to watching the news, and it could be because the news gives us the exact ratio we're seeking subconsciously. This is why we tune into these stations and give them our focus and attention. Maybe the news stations aren't just *reporting* the news but rather *projecting* it. Maybe they're projecting alternate realities to us that we in turn manifest because we became the observer. That's something to consider.

Not only are 80% of our thoughts negative in nature, but studies also show that 95% of our daily thoughts are the same thoughts we had the day before. Meaning, we're caught on a hamster wheel and keep recycling the same crappy thoughts every day. It sounds like Einstein's definition of insanity, doing the same thing over and over and expecting a different result.

Centuries ago, we woke up and worried about hunting for our food and escaping death from saber-toothed tigers that stalked us like cattle. We asked every day, "How do I survive today?"

Today, it's different. We don't have those same problems, yet we still have the same primitive thinking of lack, fear, and scarcity. We need to upgrade our thinking process and ask better questions like, "How do I enjoy my day today? How can I live in my purpose, and how do I live my best life?" Yet, we're still asking the question, "How do I survive?" There's a study that shows when people get out of *survival mode* thinking, their productivity, creativity, and overall health and well-being increase tremendously.

We can thrive in life instead of just trying to survive. We live in a world with an abundance of everything we need, and there's no reason for people to continue living in primitive mindsets any longer.

We keep returning to one primary truth: that what we focus our power of attention on is what we generate in the world. Each moment, each of us controls where we focus our almighty attention.

We must assume full responsibility each moment for where we're aiming our attentive power if we want to change the world.

This is the biggest leap we must make as we evolve into a fully-empowered quantum world view. It's encapsulated in the saying:

> *"If you want to change the world*
> *first of all you must act*
> *to change yourself."*

TRANSFORMING GLOBAL BELIEFS

Carl Jung was the first psychologist to advocate for the concept of a global consciousness; whatever the masses believe in will tend to manifest. So how do we as individuals change a global belief, such as going from scarcity to abundance, from war to peace, from competition to cooperation, or even from hate to love?

I'm reminded of the old, universal belief that no one could break the four-minute mile in a race. The ancient Romans would try to get their curriers (runners) to go faster and faster, but there seemed to be a limit to how fast the human body could run, even with the best training and discipline. This belief was propagated over and over for centuries, until finally a disbeliever in the runners' speed limit, Roger Bannister, broke the four-minute mile barrier in 1956.

How he accomplished this seemingly-impossible feat is worth noting. What he did first, as is well-documented, was mentally break his mind free from the pervading belief that no one could run beyond a certain speed. Then he was able to act and accomplish this goal physically. He celebrated winning before he won, he threw a big party before the race, and regularly visualized himself accomplishing the goal. He programmed his subconscious mind with a new positive belief in his physical powers.

For many decades, a great many runners had tried to break the four-minute barrier, and all of them had failed, even with great physical training. Then Bannister disproved the ingrained belief that the four-minute mile was impossible, just a week afterward, someone else was able to do the same, after the old belief had been shattered. Runners from all over the world started breaking the four-minute mile, including even high school students.

The only thing stopping people before seems to have been the global belief that it couldn't be done. Everyone accepted the general consensus that it was impossible, the collective consciousness believed it wasn't possible, so that made it not possible, until just one person stepped outside global consciousness with his own imagination, willpower, and positive belief. He acted as if it had already happened, and ... it happened.

That's how global change happens; one person or a small group challenges an ingrained belief, such as the world is flat, for instance, or that we can't fly to the moon. Once that person or group shows the world that the new belief is valid, *then* that accomplishment is downloaded into the collective consciousness.

Our global or societal beliefs hold us back from evolving. Breakthroughs are held back for so long, until someone pushes that bubble enough and it breaks. This is often referred to as the *tipping point*, where one person or a group transcends an old assumption or belief, then the population as a whole also shifts and goes for it.

Imagine what would happen if we could gather a large group of people who believed they could change something in the world. What if we could heal the world of its violence, hatred, and scarcity? Imagine what would happen if we consciously decided to unite the new quantum insights outlined in this book with our wide-spread humanitarian goals. I firmly believe this is where quantum science and human evolution can become entangled for the common good.

Let's finish by focusing on how you feel about your role and responsibility in generating the change you want to see happen in the world. Do you feel that you as an individual can have any impact on solving the massive negative problems on our planet?

You've learned or been reminded of many quantum facts and realizations, showing that you're much more than you were initially taught to believe—and that you pack much more manifestation power than was previously believed. But ... what are you going to do about it? Being inspired is fun, but how does an initial inspiration translate into action and transformation?

As always, the answer to such deep questions is found not just through pure deductive reasoning and intellectual assertion. The answer is found when you tune into your breathing and your feelings, your heartfelt emotions and yearnings. As your breaths now come and go, all on their own, and you relax and open up to listen to the inner voice of your heart speaking wisdom to you ... just notice whatever thoughts and feelings come to mind.

LEAPING INTO GREATNESS AND FULFILLMENT

The very idea that you can actively transform your mindset in unseen quantum directions is a somewhat radical proposition. For countless generations, people all over the world lived their lives within a stable world view that changed very little from childhood to old age. Yet technology and science has changed all that. We're now suddenly riding an accelerating arc of material change that in turn demands radical change in our minds.

When we talk about mindset, we're referring to our entire mental and emotional conditioning from birth onward, plus the deeper inherited aspects as well. Our mindset consists of all our beliefs, conscious and unconscious, all our programmings and assumptions about reality. Our mindset also determines how we think, where we focus our attention, and how we manage our own awareness. It includes our ego sense of who we are, our root identity and sense of individuality.

Humanity as a whole is being challenged to our evolutionary limits right now. We're all trying to maintain inner balance while doing our best to embrace new scientific concepts that at first seemed bizarre, fanciful, or downright irrational. Hopefully, as we open up to the emerging quantum vision of reality, we will find when we let go of old Newtonian models and entertain a bright new vision of how things work, we can relate with our universe much better than before.

To sum it all up: As we begin to comprehend reality at a higher level, we're finding that we can manage the flow of our lives more effectively. We can stay open to change, detect previously-unseen options, and respond to new situations with more clarity, confidence, and successful responses. As anxiety about the unknown reduces and life becomes to make more sense, we're also finding that we can relax within our expanded world view and enjoy a heightened sense of fulfillment.

Most importantly, we're beginning to shift out of old manipulatory habits where we saw ourselves as separate isolated entities fighting against the world. Instead, we're embracing a participatory mindset where we can feel our engagement in all of life in a new way.

Quantum physics offers us both new scientific vistas, and also new (and ancient) metaphors about how we fit into a much larger, seemingly-infinite reality where everything is related to everything else. Perhaps most importantly, we're finally coming to realize that even in science, there exists a dimension of consciousness that seems to transcend the spacetime continuum, integrating science and spirit into a truly universal and harmonious whole.

Einstein himself believed in a spiritual underpinning of physical reality. Now, with string theory, we're exploring the scientific validity of such things as multiple dimensions, even multiple universes. We now know that sudden quantum leaps of electrons from one orbit to another seem to happen instantaneously, with no passage of time. We also know two entangled particles at great distances from each other can respond to change instantly.

We're also coming to accept that quantum entanglement of quarks and gluons, photons and electrons, protons and neutrons and all the rest seem to be what generates all the complexity found in the universe. Energetic entanglement could also explain such phenomena as ESP (Extrasensory Perception), where people's minds seem to connect and communicate from even halfway around the globe.

The notion that consciousness exists outside the bounds of space and time is radical and requires a quantum leap to even consider that our thoughts, feelings, and inner conscious experience is functioning beyond the laws of classical physics. Once we entertain this new belief, this expanded model of reality, we find ourselves empowered to tap into new realms of thought and action.

We're transforming our old beliefs of who we truly are, and in the process, we're coming to realize that we're all intimately and instantaneously connected on subtle levels and with the whole universe. A profound sense of unity emerges, as does a remarkably-good feeling in our hearts.

This "lovingly-entangled" feeling may explain the universal power in Jesus' equation that "God is love." Seen from a quantum mindset, we're all participating in this infinite resonant unified reality. Our challenge is to entertain the belief and see if it feels valid and rings true in our hearts and minds. If it does, we need to leap into active expression of all that it implies.

Right now, having read this book and opened yourself to your own experience of mindset expansion, you're ready to sense your engagement with everyone and everything around you in a new way. Like particles that are energetically entangled with each other, you can feel your loving connection both with people you know, and ultimately with the whole of humanity.

Perhaps this energetic entanglement explains how two people, never having met before in their lives, can encounter each other and instantly feel some kind of intimacy, some subtle entanglement of their feelings and thoughts. Love at first sight, where there's instantly a deep sense of engagement and communication, is quite possibly a natural function of quantum entanglement.

At the quantum level, we are all immersed in and interacting through one unifying and vastly intelligent energy field. Research has now shown how, at subtle levels that are usually mostly unconscious, we're all resonating together within what Carl Jung called the collective unconscious. This heartfelt sense of shared resonance and participation is hopefully becoming more and more conscious with the dawning of each new day.

Of course, we need to maintain our ego's illusion that we're separate individuals. Our sensory inputs of seeing and hearing keep telling us that our body is a separate entity fixed in local space and time. But at the quantum level we're realizing, as Buddha, Lao Tzu, Jesus, and all the other deep mystics have insisted, that we're all one. Each person's thoughts and actions affect the next person.

The quantum mindset challenges us to enhance our world view and leap into this expanded understanding of who we truly are—conscious individuals inextricably entwined with each other and the earth, and even with the stars. These new quantum models and discoveries are often a challenge to grasp, to explore, and to ultimately embrace, but the evolution of human consciousness seems to be guiding all of us in this more-awakened direction. We're transcending limiting biological perceptions and sensory assumptions and leaping into a higher sense of how we're engaged with ourselves, our loved ones, and the whole world.

REMEMBER TO EXPECT MIRACLES!

On your journey of co-creation and manifestation, you will discover many synchronicities and serendipitous moments where it either seems too good to be true or too much of a coincidence. This is never the case. All things in your life happen for you, not to you. When magic happens in your life, it's normal, because it's the natural eb and flow of the workings of this Universe.

When these moments happen, don't develop new negative one-liners like, "That's impossible," "It can't be that easy," or, "This has to be a coincidence." It will *feel* like it's magic but get over it. It's hard to accept it's easy, but it is. In fact, life is so easy that it's hard. We're the ones who make it harder than it is, usually by overthinking and over-complicating things.

You are an amazing co-creator in this universe, and the universe not only allows for your participation but requires it. Now go out into the world and expect miracles. Watch how fast your life changes for the greater good.

FINAL WORDS:

I'll end with a story about a man who steals an egg from an eagle's nest and puts it in the nest of a chicken, where it hatches and grows up as a chicken. All its senses indicate it's a chicken, so it learns to act like a chicken. One day, the eagle momma flies down and snatches the young eagle up and brings it back to its rightful nest, but the eagle doesn't think it can fly. It's been so programmed by its chicken family that it doesn't believe it can soar into the air like its true momma and eagle siblings.

The little eagle stays high up in the nest, afraid to take the leap into the great beyond like its siblings have already. Suddenly one day, it is pushed out of the nest. It falls and falls, and only at the last moment its true instinctive nature takes over and over-rides its programming. The eagle's spirit and whole being finally wake up to its true potential, it flaps its wings, and flies high!

Maybe our unconscious negative and limited-belief programing is the reason why many people feel like they're cruising through life with one foot on the gas and one foot on the brake yet can't understand why it's so hard to reach their destination.

Like the eagle, you're far more powerful than you're aware. You playing small does not serve you, your family, or the world at large. Hopefully this book will be the little "nudge" out of your comfort zone that you need to take a quantum leap into a new way of thinking. When you change your thinking, you change your world and the world around you. Just have faith and remember that sometimes you have to jump to find your wings on the way down.

Feeling uncomfortable is a sign of growth and new beginnings, so embrace it. Soon you'll see the universe has your back and manifesting is much easier than you thought. When you find comfort in your discomfort and accept that anything is possible—you'll be unstoppable!

To infinity and beyond...

ABOUT THE AUTHOR

Rick Thompson is an engineer, businessman, and investor based in Seattle, Washington. He is a CEO in the cement and steel industry and is recognized as a leading authority on post-tensioning procedures working with big-tech clients such as Microsoft and Google.

Rick has been infatuated with science, math, and the workings of the world since childhood. He is highly skilled at defining objectives, assessing requirements, and resolving problems, in and out of the work field. His practical approach to business, life, and science is what makes this book special and very digestible for the average reader.

This book and guided programs will provide you with a tested nuts-and-bolts method for entering the quantum mindset and tapping heightened tools for manifesting your deepest dreams.